BARRON'S
NEW YORK STATE

GRADE 8 MATH TEST

THIRD EDITION

Anne M. Szczesny, M.Ed.
Gifted Math Program
University at Buffalo

All inquiries should be addressed to:
Barron's Educational Series, Inc.
250 Wireless Boulevard
Hauppauge, New York 11788
www.barronseduc.com

ISBN: 978-0-7641-4623-7

ISSN: 1534-326X

PRINTED IN THE UNITED STATES OF AMERICA
9 8 7 6 5 4

10%
POST-CONSUMER
WASTE
Paper contains a minimum of 10% post-consumer waste (PCW). Paper used in this book was derived from certified, sustainable forestlands.

Table of Contents

Chapter 1

The Test

The New York State Grade 8 Math Test is used to see how well you
- listen to and follow directions
- have mastered mathematical concepts
- reason when solving real-world problems
- use and apply formulas, definitions, signs, and symbols
- read and interpret tables and graphs
- choose and use the appropriate procedure in solving a problem.

Beginning in May 2011, eighth-grade students are tested on material learned in May and June of grade 7 and on material taught in September through April of grade 8.

The timed test consists of two parts. Part I has 42 multiple-choice questions. You have 65 minutes to answer those questions. In Part 2, you will get 70 minutes to answer 8 short- and 4 extended-response questions.

Each multiple-choice question has four answers from which to choose the correct one. For extended-response questions, you must provide an answer as well as show how you arrived at your answer. There is no *one* correct explanation for the extended-response questions. Be clear in your explanation. You can earn partial credit on these problems.

You are *not* allowed to use a calculator for the multiple-choice questions. However, you will need a scientific calculator to solve the short- and extended-response questions. If you don't have a scientific calculator, use a four-function calculator ($+, -, \times, \div$) with a square-root key ($\sqrt{}$). You will also need a ruler divided into sixteenths. A reference sheet with some advanced formulas will be provided.

SCORING

Each correct multiple-choice question is graded as either right or wrong. The extended-response questions are graded using a rubric, which is a set of guidelines teachers follow to award points. You can earn 0, 1, or 2 points on the short-response questions and 0, 1, 2, or 3 points on the extended-response questions.

The test results will be reported on a scale of 1 to 4, where 4 is the highest. These results are used by your teachers to see where you may need more study or practice in math.

TEST-TAKING TIPS

Use the following test-taking tips to help you score high on the math test.

■ Relax! You've taken timed tests before. This test gives you the opportunity to show your teachers how savvy you are in math!
■ Listen carefully to and follow all the directions given by your teacher.
■ Answer all the questions that seem easy to you. Then go back and work on the other questions for which you need more time.
■ Make sure you read the problem carefully and understand what the question is asking of you.
■ Use the problem-solving techniques that you have learned, such as working backward, making lists, drawing diagrams, guessing and testing, and breaking a large problem into smaller ones. When you get the answer, label it correctly with the proper measurements

or units. For example, if you are answering a question that asks for the area of a triangle, label your answer with square units, such as square inches or square meters.

▪ If you have extra time, go over all your answers to make sure they are correct. Write neatly because other teachers besides your math teacher will be scoring the tests.

▪ Do not leave any blanks! They will be scored as incorrect, and you will receive zero points. Take a guess if you are not sure of an answer.

SAMPLE QUESTIONS

Here are samples of the types of questions you will be answering on the test.

Multiple-Choice Question

1. Two angles are complementary. One angle measures 17°. What is the measure of the other angle?
(A) 163°   (B) 78°   (C) 73°   (D) 117°

Solution

The correct answer is (C). The sum of complementary angles is 90° (17° + 73° = 90°).

Short-Response Question

2. Use the chart below to answer the questions.

x	y
1	
5	1
10	2
	3
20	4

A. Fill in the two missing values.

B. Write the equation that represents the relationship between x and y.

Solution

A. The missing y-value is $\frac{1}{5}$, or 0.2, and the missing x-value is 15.

B. The value of x is 5 times the value of y. The equation can be written as $x = 5y$, or $y = \frac{x}{5}$.

A point value of 2, 1, or 0 can be awarded for this type of question.

Extended-Response Question

3. To control her blood pressure, Jill's grandmother takes one-half of a pill every other day. One supply of medicine contains 60 pills. Approximately how many months will one supply of pills last? Explain your answer.

Solution

Taking one-half of a pill every other day is another way of saying that Jill's grandmother takes 1 pill every 4 days. At this rate, 60 pills will last 240 days. There are approximately 30 days in a month, and 240 days is roughly 8 months.

So, one supply of pills will last approximately 8 months.

A point value of 3, 2, 1, or 0 can be awarded for answers to longer extended-response questions.

This book was written to help you review the math topics you have forgotten or need to practice. Practice questions are at the end of each chapter. Solutions with explanations are provided, but you may find another solution that is also acceptable. The questions are similar to the ones that appear on the test. Grab some paper, a few pencils, and your calculator and work along with the examples throughout each chapter. In addition to the practice questions, two sample tests are included for practice.

More information about the testing program, including a glossary of terms and sample test questions, can be found on the New York State Education Department web site at www.p12.nysed.gov/osa/ei/eigen.html.

Follow the examples in this book. Practice answering the *Test Your Skills* questions and completing the sample tests. If you don't understand something, ask your teachers. They are there to help you. Good luck!

Chapter 2

Numbers and Operations

Note: The material presented in this chapter is a review of the concepts on which your present math skills are based. This material will not be on the test as it is presented here. It is included to refresh your memory.

NUMBER SETS

The first set of numbers people used to count was the set of *natural numbers*, or *counting numbers*.

$$N = \{ 1, 2, 3, 4, 5, ...\}$$

To this set we added zero and formed the set of *whole numbers*.

$$W = \{ 0, 1, 2, 3, 4, ...\}$$

Negative numbers, or additive inverses, were added to the set of whole numbers to form the set of *integers*. The first letter in the word *Zahlen*, the German word for number, represents this set.

$$Z = \{ ..., -3, -2, -1, 0, 1, 2, 3, ...\}$$

To represent parts of a whole, rational numbers are used. We use set builder notation to describe the elements of this set, which includes any number that can be written as a ratio. The letter Q, the first letter in the word *quotient,* represents the set of rational numbers.

$$Q = \{ q \mid q = \frac{a}{b}, \text{ where } a, b \text{ are integers and } b \neq 0 \}$$

We read this as "The set **Q** contains all elements q such that $q = \frac{a}{b}$, where a and b are integers and b is not equal to 0."

The *irrational numbers* are those numbers that cannot be written as a ratio of two integers. Examples of irrational numbers are π, $\sqrt{2}$, and numbers such as 4.375728..., where the digits to the right of the decimal point do not repeat in any sort of pattern. When finding $\sqrt{2}$, your calculator display shows the decimal number 1.41421356..., whose digits do not follow a pattern. Not all square roots are irrational numbers. For example, $\sqrt{36} = \pm 6$ since $6^2 = 36$ and $(-6)^2 = 36$. If the number under the radical sign is a perfect square, the number is rational.

A *Venn diagram* is used to show how the sets of numbers are related to each other. Together, the rational numbers joined with the irrational numbers compose the set of *real numbers,* R. In the diagram, the size of the ovals does not reflect the size of the set of numbers.

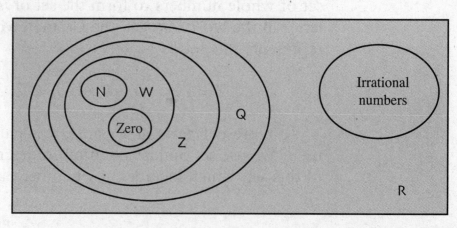

PROPERTIES OF OPERATIONS

The operations of addition and multiplication are *commutative*. When adding or multiplying two numbers, the order of the numbers does not affect the answer.

$$a + b = b + a \qquad c \cdot d = d \cdot c$$
$$3 + 7 = 7 + 3 \qquad 6 \cdot 2 = 2 \cdot 6$$
$$10 = 10 \qquad 12 = 12$$

The operations of addition and multiplication are *associative*. Different groupings of the numbers do not affect the answer.

$$a + (b + c) = (a + b) + c \qquad f \cdot (g \cdot h) = (f \cdot g) \cdot h$$
$$3 + (9 + 4) = (3 + 9) + 4 \qquad 4 \cdot (5 \cdot 2) = (4 \cdot 5) \cdot 2$$
$$3 + 13 = 12 + 4 \qquad 4 \cdot 10 = 20 \cdot 2$$
$$16 = 16 \qquad 40 = 40$$

The operation of multiplication is *distributive over addition*.

$$a \cdot (b + c) = (a \cdot b) + (a \cdot c)$$
$$2 \cdot (5 + 3) = (2 \cdot 5) + (2 \cdot 3)$$
$$2 \cdot 8 = 10 + 6$$
$$16 = 16$$

A number is called the *additive inverse* of a given number when the sum of the two numbers is zero. Here a is also called the additive inverse of $-a$, and $-a$ is called the additive inverse of a. Numbers are opposites if they are the same distance away from zero on the number line.

$$a + (-a) = 0$$

The *multiplicative inverse* (or *reciprocal*) of a number multiplied by the given number always equals 1.

$$b \cdot \frac{1}{b} = 1 \quad \text{and} \quad -b \cdot \frac{1}{-b} = 1$$

The *zero property for addition* states that when zero is added to any number, the result is the original number.

$$c + 0 = c \text{ and } 0 + c = c$$

The *identity element for multiplication* is 1. When a number is multiplied by 1, the product is the original number.

$$d \cdot 1 = d \text{ and } 1 \cdot d = d$$

The *zero property of multiplication* states that any number multiplied by zero equals zero.

$$g \cdot 0 = 0 \text{ and } 0 \cdot g = 0$$

ABSOLUTE VALUE

The absolute value of a number is its distance away from zero on the number line. The symbol for absolute value is a pair of vertical lines.

EXAMPLES

A. $|+10| = 10$

B. $|24| = 24$

C. $|-16| = 16$

D. $\left| \dfrac{-5}{13} \right| = \dfrac{5}{13}$

Find the absolute value of a number to get its *magnitude*. The magnitude of −36 is 36. The magnitude of +19 is 19.

OPERATIONS ON INTEGERS

Adding Integers

Positive + positive $\Rightarrow$ positive $(7 + 4 = 11)$
Negative + negative $\Rightarrow$ negative $(-4 + -17 = -21)$

When the signs on the numbers are different, subtract the absolute value of the numbers and keep the sign of the number with the larger magnitude in your answer.

EXAMPLES

    A. $-16 + 4 = -12$
    B. $-2 + 7 = 5$

Subtracting Integers

To subtract signed numbers, rewrite the problem as an addition problem by adding the inverse of the second number.

EXAMPLES

    A. $15 - 6 =$
         $15 + (-6) = 9$

    B. $-7 - 4 =$
         $-7 + (-4) = -11$

    C. $3 - 9 =$
         $3 + (-9) = -6$

    D. $29 - (-7) =$
         $29 + (+7) = 36$

    E. $-12 - (-20) =$
         $-12 + (+20) = 8$

Multiplying Integers

Positive • positive ⇒ positive
Negative • negative ⇒ positive
Negative • positive ⇒ negative
Positive • negative ⇒ negative

EXAMPLES

A. $7 \cdot 8 = 56$ (There are 7 boxes of crayons with 8 crayons in each.)

B. $-6 \cdot 8 = -48$ (I owe 6 cents to each of 8 friends.)

C. $6 \cdot -4 = -24$ (Over the last 6 months, my locker rental bill was $4 per month.)

D. $-6 \cdot -3 = 18$ (If you went on a diet and *lost* 3 pounds per month, 6 months *ago* you were 18 pounds heavier.)

Dividing Integers

$\dfrac{positive}{positive} \Rightarrow positive \left(\dfrac{72}{9} = 8 \right)$ $\dfrac{positive}{negative} \Rightarrow negative \left(\dfrac{126}{-9} = -14 \right)$

$\dfrac{negative}{negative} \Rightarrow positive \left(\dfrac{-63}{-7} = 9 \right)$ $\dfrac{negative}{positive} \Rightarrow negative \left(\dfrac{-54}{3} = -18 \right)$

EXAMPLES

A. $\dfrac{24}{3} = 8$ (Placing 24 cupcakes on 3 plates means there are 8 per plate.)

B. $\dfrac{15}{-5} = -3$ (At the video arcade, you had $15 more 5 hours ago. Your spending rate is $3 per hour.)

C. $\dfrac{-35}{7} = -5$ (Spending $35 at an amusement park over 7 hours is the same as spending an average of $5 per hour.)

D. $\dfrac{-42}{-6} = 7$ (If a person wants to lose 42 pounds and plans to lose 6 pounds per month, it should take 7 months.)

ORDER OF OPERATIONS

What is the value of x in the equation $15 + 9 \div 3 = x$? To be sure that everyone gets the same answer, an order of operations was developed. This is the order in which we perform operations.

1. Parentheses—work inside the grouping symbols
2. Exponents—simplify the terms with exponents
3. Multiplication / division—performed from left to right
4. Addition / subtraction—performed from left to right

Scientific calculators perform operations according to this order.

When not using a scientific calculator, the mnemonic (memory aid) "Please excuse my dear Aunt Sally" will help you remember the correct order of operations.

EXAMPLES

Evaluate these expressions.

A. $15 + 9 \div 3$
$15 + 3 = 18$

B. $3^2 + (4 - 7) \div 3$
$3^2 + (-3) \quad \div 3 =$
$9 - \quad 3 \quad \div 3 =$
$9 - \quad 1 \qquad = 8$

C. $18 \div (7 - 4^2) \cdot 7$
$18 \div (7 - 16) \cdot 7 =$
$18 \div (-9) \quad \cdot 7 =$
$-2 \qquad \cdot 7 = -14$

ADDING AND SUBTRACTING FRACTIONS

Add or subtract the numerators when the denominators are the same. Keep the same denominator. Reduce the fraction to lowest terms if needed.

EXAMPLES

A. $\dfrac{5}{12} + \dfrac{1}{12} =$

$\dfrac{5+1}{12} = \dfrac{6}{12} = \dfrac{1}{2}$

B. $\dfrac{19}{27} - \dfrac{16}{27} =$

$\dfrac{19-16}{27} = \dfrac{3}{27} = \dfrac{1}{9}$

C.

$$\begin{array}{r} 5\dfrac{3}{7} \\[2mm] +\,6\dfrac{6}{7} \\ \hline 11\dfrac{9}{7} = 12\dfrac{2}{7} \end{array}$$

D.

$$\begin{array}{r} 2\dfrac{3}{8} \\[2mm] -\,1\dfrac{1}{8} \\ \hline 1\dfrac{2}{8} = 1\dfrac{1}{4} \end{array}$$

E.

$$7\frac{3}{11} = 6\frac{14}{11}$$

$$-3\frac{6}{11} = 3\frac{6}{11}$$

$$3\frac{8}{11}$$

You cannot subtract $\frac{6}{11}$ from $\frac{3}{11}$, so rename $7\frac{3}{11}$ as $6\frac{14}{11}$ and subtract as usual.

When fractions have different denominators, find the least common denominator (LCD) and then rewrite equivalent fractions using the LCD.

EXAMPLES

F.

$$\frac{5}{12} + \frac{1}{4} =$$

$$\frac{5}{12} + \frac{3}{12} =$$

$$\frac{5+3}{12} = \frac{8}{12} = \frac{8 \div 4}{12 \div 4} = \frac{2}{3}$$

G. $\dfrac{3}{4} + \dfrac{1}{5} + \dfrac{1}{2} =$

$$\dfrac{15}{20} + \dfrac{4}{20} + \dfrac{10}{20} = \dfrac{29}{20} = 1\dfrac{9}{20}$$

H. $\dfrac{1}{2} - \dfrac{3}{7} =$

$$\dfrac{7}{14} - \dfrac{6}{14} = \dfrac{1}{14}$$

I.

$$2\dfrac{1}{8} = 2\dfrac{3}{24}$$

$$+1\dfrac{2}{3} = 1\dfrac{16}{24}$$

$$\overline{3\dfrac{19}{24}}$$

J.

$$5\dfrac{1}{7} = 5\dfrac{6}{42} = 4\dfrac{48}{42}$$

$$-2\dfrac{5}{6} = 2\dfrac{35}{42} = 2\dfrac{35}{42}$$

$$\overline{2\dfrac{13}{42}}$$

MULTIPLYING AND DIVIDING FRACTIONS

The product of fractions is obtained by multiplying the numerators together to get a new numerator and then multiplying the denominators together to get a new denominator. Change mixed numbers to improper fractions and then multiply.

EXAMPLES

A. $\dfrac{2}{9} \cdot \dfrac{3}{4} \cdot \dfrac{2}{3} = \dfrac{\cancel{2}}{9} \cdot \dfrac{\cancel{3}}{\cancel{4}} \cdot \dfrac{\cancel{2}}{\cancel{3}} = \dfrac{1}{9}$

B. $6\dfrac{1}{5} \cdot 4\dfrac{3}{4} =$

$\dfrac{31}{5} \cdot \dfrac{19}{4} = \dfrac{589}{20} = 29\dfrac{9}{20}$

To divide fractions, change the divisor (the number after the division sign) to its reciprocal and then multiply. To find the reciprocal of a mixed number, change the mixed number to an improper fraction and then change the improper fraction to its reciprocal. The reciprocal of

$2\dfrac{3}{5}$ is $\dfrac{5}{13}$.

$2\dfrac{3}{5} = \dfrac{13}{5}$ and then invert $\dfrac{13}{5}$ to $\dfrac{5}{13}$

EXAMPLES

C. $\dfrac{3}{7} \div \dfrac{6}{11} =$

$\dfrac{3}{7} \cdot \dfrac{11}{6} = \dfrac{33}{42} = \dfrac{33 \div 3}{42 \div 3} = \dfrac{11}{14}$

D. $3\dfrac{1}{5} \div 4 =$

$\dfrac{16}{5} \cdot \dfrac{1}{4} = \dfrac{16}{20} = \dfrac{16 \div 4}{20 \div 4} = \dfrac{4}{5}$

E. $7 \div 2\frac{1}{3} =$

$$\frac{7}{1} \cdot \frac{3}{7} = \frac{21}{7} = \frac{21 \div 7}{7 \div 7} = \frac{3}{1} = 3$$

DIVISIBILITY RULES

A number is divisible by

2 if the number is even or the last digit is divisible by 2.

3 if the sum of the digits in the number is divisible by 3.

4 if the last two digits are divisible by 4.

5 if the last digit is 5 or 0.

6 if the number is even <u>and</u> the sum of the digits is divisible by 3.

8 if the last three digits are divisible by 8.

9 if the sum of the digits is divisible by 9.

10 if the last digit is 0.

GREATEST COMMON FACTOR AND LEAST COMMON MULTIPLE

When given two or more numbers, the greatest common factor (GCF) is the largest factor all the numbers share.

EXAMPLE

Find the GCF of 20 and 48.
Write out the factors of each number. Identify the largest factor common to both 20 and 48.

20: 1, 2, ④, 5, 10, 20
48: 1, 2, 3, ④, 6, 8, 12, 16, 24, 48

Thus, the GCF of 20 and 48 is 4.

The least common multiple (LCM) is the smallest multiple the numbers have in common.

EXAMPLE

Find the LCM of 20 and 48.
Write out multiples of the numbers, and find the smallest number that is in both lists.

20: 20, 40, 60, 80, 100, 120, 140, 160, 180, 200, 220, ②④⓪, 260, ...
48: 48, 96, 144, 192, ②④⓪, 288, 336, ...

Therefore, the LCM of 20 and 48 is 240.

PRIME NUMBERS, COMPOSITE NUMBERS, AND EXPONENTS

A *prime number* is a whole number greater than 1 that has exactly two different factors, itself and 1. The number 1 is *not* prime because it has only one factor, 1. The only prime number that is even is 2. Here is a set of the first 12 prime numbers.

$$\{2, 3, 5, 7, 11, 13, 17, 19, 23, 29, 31, 37\}$$

If a number has more than two factors, it is a *composite number*. Composite numbers can be written as a product of prime factors.

$$56 = 2 \cdot 2 \cdot 2 \cdot 7$$

When the same factor is used more than once to obtain a product, we use exponents. For example, $81 = 3 \cdot 3 \cdot 3 \cdot 3 = 3^4$. The *exponent* is 4, and the *base* is 3. If there is no exponent on a factor, the factor is used once. The number 1 is *never* written as an exponent on a factor.

$$56 = 2 \cdot 2 \cdot 2 \cdot 7 = 2^3 \cdot 7$$

SCIENTIFIC NOTATION

Very small and very large numbers can be written in scientific notation as a number greater than or equal to 1 and less than 10 multiplied by a power of 10. For example, 3,570,000,000 can be written as $3.57 \cdot 10^9$. The number 0.000000719 can be written as $7.19 \cdot 10^{-7}$.

The distance between the sun and Earth is roughly 93,000,000 miles, or $9.3 \cdot 10^7$ miles. A micron is one-millionth of a meter. Most cells in your body have a diameter of 5 microns (or $5 \cdot 10^{-6}$ meters) to 12 microns (or $1.2 \cdot 10^{-7}$ meters).

TEST YOUR SKILLS   (See page 159 for answers.)

1. $4 + 3(15 - 3) \div 6 + 2^3 =$
   (A) 22   (B) 18   (C) $\dfrac{5}{64}$ (D) 6

2. Which operation should you perform first in solving this equation?
   $36 + 9 \div 5 \cdot 3 - 2 = k$
   (A) +   (B) ÷   (C) –   (D) •

3. The reciprocal of $7\dfrac{1}{3}$ is

   (A) $\dfrac{8}{3}$ (B) $\dfrac{3}{11}$ (C) $\dfrac{22}{3}$ (D) $\dfrac{3}{22}$

4. Between which two integers does $\sqrt{152}$ lie?
   (A) 12, 13   (C) 144, 169
   (B) 151, 153   (D) 14, 16

5. Ms. Peters asked her students to substitute 3 for x in the expression $4x^2$. Dave said the expression was equal to 36, but Jill said it was equal to 144. Who is correct and why?

6. The GCF of 16 and 64 is
   (A) 64   (B) 8   (C) 16   (D) 4

7. How many pieces of ribbon can be cut from a 20-inch piece if each piece has to be $2\dfrac{1}{2}$ inches long?

8. When written in scientific notation, the product of $7.8 \cdot 10^5$ and $6.2 \cdot 10^3$ is
   (A) $14 \cdot 10^8$ (C) $48.36 \cdot 10^8$
   (B) $4.836 \cdot 10^8$ (D) $4.836 \cdot 10^9$

9. Use the distributive property to simplify $\dfrac{4}{3}(6+9)$. Show the steps used.

10. $12 \cdot 3\dfrac{5}{6} = ?$

Chapter 3

Patterns and Functions

WRITING EXPRESSIONS

A pattern is a model developed after observing examples. Find a rule or algebraic expression that will give you the next three numbers in the sequence below.

$$3, 10, 17, 24, 31, 38, \underline{\quad}, \underline{\quad}, \underline{\quad}, \ldots$$

To get the next number, 7 is added to the previous number. Add 7 to 3 to get 10, 7 to 10 to get 17, and $17 + 7 = 24$. The three missing numbers are 45, 52, and 59. We can represent the pattern algebraically using the expression $n + 7$.

EXAMPLES

Write an algebraic expression that describes the pattern of these sequences, and use it to find the next three numbers.

A. $-6, -11, -16, -21, \underline{\quad}, \underline{\quad}, \underline{\quad}, \ldots$
$n - 5$ gives you $-26, -31, -36$.

B. $729, 243, 81, \underline{\quad}, \underline{\quad}, \underline{\quad}, \ldots$
$\dfrac{n}{3}$ gives you $27, 9, 3$.

C. $-3, -6, -12, -24, \underline{\quad}, \underline{\quad}, \underline{\quad}, \ldots$
$2n$ gives you $-48, -96, -192$.

Mathematics is a language, just like English. You often have to translate between the two. Mathematical expressions can often be written in more than one way. The expression $2n + 1$ can be translated into English as *double the number and add 1* or as *add 1 to twice a number*. The phrase *divide the number by 4 and then subtract 7* can be translated into the algebraic expression $\frac{n}{4} - 7$.

EXAMPLES

Write an English phrase for each of the following.

D. $23 - 6$
Twenty three minus 6, 23 diminished by 6, or 6 less than 23

E. $5 + 3x$
Five plus 3 times a number, 5 more than a number tripled, or 3 times a number increased by 5

F. $7n - 3$
Seven times a number diminished by 3, 3 less than 7 times a number

G. $20 - 4n$
Twenty minus 4 times a number, 4 times a number subtracted from 20, or 20 diminished by 4 times a number

H. $\frac{2n - 3}{7}$
The quantity 3 less than twice a number is divided by 7, or the quotient of twice a number diminished by 3 then divided by 7

I. $4(7 + 5n)$
The product of 4 and the quantity 7 plus 5 times a number, or the sum of 7 and 5 times a number is then multiplied by 4

Write the algebraic expression for the following examples.

J. Tomas is 7 years older than his sister. Express Tomas's age with respect to his sister's age.

Let s be his sister's age. Tomas's age is $s + 7$ or $7 + s$.

K. Cara's test score is 10 points higher than Jen's. Express Jen's score in relation to Cara's.

If Cara's score is 10 points higher than Jen's, then Jen's score is 10 points lower than Cara's. So Jen's score is represented by $c - 10$.

L. Jill has one more than twice the number of pencils Tony has. Represent the number of pencils Jill has with respect to Tony's.

Jill has $2p + 1$ pencils, where p is the number of pencils Tony has.

EVALUATING EXPRESSIONS

The value of the variable for substitution is given when you are asked to evaluate an expression. To evaluate the expression $6d - 5$ when d is 2, substitute 2 for d.

$$6d - 5$$
$$6(2) - 5 =$$
$$12 - 5 = 7$$

EXAMPLES

Evaluate the following expressions.

A. $7 - 3x$, where $x = 6$
$7 - (3)6 =$
$7 - 18 = -11$

B. $30 - (9 - 4y)$, where $y = -3$
$30 - (9 - 4(-3)) =$
$30 - (9 + 12) =$
$30 - 21 = 9$

C. $\dfrac{2k-3}{7}$, where $k = 19$

$$\dfrac{2(19)-3}{7} =$$

$$\dfrac{38-3}{7} = \dfrac{35}{7} = 5$$

WRITING EXPRESSIONS AND EQUATIONS FROM DATA

The sequence 1, 4, 9, 16, ____, ____, ____, ... contains perfect squares. The first number is 1^2, the second number is 2^2, and the third number is 3^2. We can put this data in a table and then fill in the empty boxes using the pattern.

n	1	2	3	4				...
n^2	1	4	9	16				...

Substitute the values of n to get n^2.

$$5^2 = 25 \qquad 6^2 = 36 \qquad 7^2 = 49$$

n	1	2	3	4	5	6	7	...
n^2	1	4	9	16	25	36	49	...

The expression that represents this pattern is n^2.

EXAMPLE

A. In the table below, the pattern is given to you. Fill in the empty boxes with the appropriate numbers. The ellipses (...) represent numbers that are not shown.

n	2	4	6	8	...			...	80
$2n + 1$	5	9			...	49	53	...	161

When $n = 6$, your answer is 13 because $2(6) + 1 = 13$. When $n = 8$, the answer is 17. To find n when your answer is 49, work backward. The last step you performed was adding 1; now subtract 1 ($49 - 1 = 48$). You multiplied by 2, so now divide by 2 ($48 \div 2 = 24$). To check, substitute 24 for n in $2n + 1$ and see if the result is 49.

n	2	4	6	8	...	24	26	...	80
$2n + 1$	5	9	13	17	...	49	53	...	161

B. Fill in the missing values, and then write an expression that describes the relationship.

x	0	1	2	3	4	5	6	...
y	1	4		10	13			...

Imagine the table as a machine. You put in 0, and 1 comes out. Put in 3, and 10 comes out. The function describes the relationship between the input and the output. In other words, it tells you what the machine did to the x-value to give you the y-value. The function of this machine is to multiply the value for x by 3 and then add 1. This result is the value y. The equation $y = 3x + 1$ describes this function. Complete the table by substituting the value of x in the equation to get the missing numbers.

$y = 3x + 1$
$y = 3(2) + 1 = 6 + 1 = 7$
$y = 3(5) + 1 = 15 + 1 = 16$
$y = 3(6) + 1 = 18 + 1 = 19$

x	0	1	2	3	4	5	6	...	n
y	1	4	7	10	13	16	19	...	$3n + 1$

C. Using the data in this table, write an equation that describes the function.

x	y
–1	–5
0	–1
1	3
2	7
3	11
4	15

Notice that the value of y is 1 less than 4 times x. The equation is $y = 4x - 1$.

D. Write an equation that describes the function from the data below.

x	y
–3	–0.75
–2	–0.5
–1	–0.25
0	0
1	0.25
2	0.5

The value of x is 4 times the value of y. The equation can be written as $x = 4y$ or as $y = \dfrac{x}{4}$.

E. Complete the table using the function $y = x^2 - 2$.

x	y
0	–2
1	
2	
3	

Substitute the values of x in the equation.

x	y
0	$y = x^2 - 2$
	$y = (-2)^2 - 2$
	$y = 4 - 2$
	$y = -2$
1	$y = (-1)^2 - 2$
	$y = 1 - 2$
	$y = -1$
2	$y = (2)^2 - 2$
	$y = 4 - 2$
	$y = 2$
3	$y = (3)^2 - 2$
	$y = 9 - 2$
	$y = 7$

The missing values are –1, 2, and 7.

EQUATIONS AND INEQUALITIES

An equation is a mathematical sentence that contains terms on both sides of an equal sign. The quantities on both sides of the equal sign have exactly the same value when the equation is true. A *term* can be a variable (x, w, or b), a constant (43, –12, or $\frac{3}{7}$), or a product of both ($5y$, $3x^3$, or b^5). The number before a variable is called the *coefficient*. In the term $2y^3$, 2 is the coefficient, y is the variable, and 3 is the exponent on y.

Equations can be true ($7 \cdot 9 = 63$), can be false ($5 - 3 = 7$), or can be open. If equations are open, they are called *open sentences*. An example of an open sentence is $2x + 7 = 15$. You cannot determine whether this equation is true or false until the variable is replaced with a number.

Inequalities are similar to equations. Instead of an equal sign, the signs < (is less than), > (is greater than), ≤ (is less than or equal to), ≥ (is greater than or equal to), or ≠ (is not equal to) are used to compare the quantities.

The *solution set* contains all the numbers that make the equation or inequality true. The solution set for $5x - 7 = 3$ is {2}, and the solution set for $2x + 7 \geq 15$ is {4, 5, 6, ...}, when choosing answers from the natural numbers.

EXAMPLE

From the set {0, 2, 4, 6, 8, 10, 12}, find all the numbers that make these statements true.

A. $t - 5 = 7$
Only the number 12 will make this statement true.

B. $13 + p \leq 21$
The solution set for this inequality is {0, 2, 4, 6, 8}.

SOLVING EQUATIONS

When solving equations, group *like terms* together. Like terms have the same variable with the same exponent. The terms $7x$ and $4x$ are like terms. The terms $3x^2$ and $9x$ are not like terms because they have different exponents. Only like terms can be combined. For example, $18x^2 + 3x^2 - 27x^2$ can be simplified to $-6x^2$ by adding the coefficients.

Regroup the terms so the terms with the variables are on one side of the equal sign and the constants are on the other. To keep your equation balanced, you must perform the same operation on both sides of the equation. Use the properties of additive inverse, multiplicative inverse (or reciprocals), commutative and associative properties, and the distributive property. Multiplying by the reciprocal is the same as dividing by the number. For example, multiplying by $\dfrac{1}{4}$ gives you the same result as dividing by 4. When you have a variable term that equals a constant, you can solve for the variable.

EXAMPLES

Solve for the variable in the following equations.

A. $2x + 7 = 21$

$$2x + 7 = 21$$
$$\underline{-7 = -7}$$
$$2x \quad = 14$$
$$\frac{2x}{2} = \frac{14}{2}$$
$$x = 7$$

B. $10n - 2 = 3n + 47$

$$10n - 2 = 3n + 47$$
$$\underline{+2 = \quad +2}$$
$$10n \quad = 3n + 49$$
$$\underline{-3n \quad = -3n}$$
$$7n \quad = 49$$
$$\frac{7n}{7} = \frac{49}{7}$$
$$n = 7$$

C. $a - (15 - a) = 45$

$$a - (15 - a) = 45$$
$$a - 15 + a = 45$$
$$2a - 15 = 45$$
$$\underline{ +15 = +15}$$
$$2a = 60$$
$$a = 30$$

D. $\dfrac{k}{5} + 8 = 43$

$$\frac{k}{5} + 8 = 43$$
$$\underline{\phantom{\frac{k}{5} +} -8 = -8}$$
$$\frac{k}{5} = 35$$
$$(5)\frac{k}{5} = 35(5)$$
$$k = 175$$

SOLVING AND GRAPHING INEQUALITIES

The procedure for solving inequalities is the same as that for solving equations. You must keep your inequality balanced. An equation usually has one solution, while an inequality may have more than one. Solutions to linear inequalities are easily graphed on the number line.

EXAMPLES

A. Graph the solution set to $x \le 4$ on the number line.

We are looking for all the numbers that are less than or equal to 4. Since 4 is to be included (notice the sign), circle 4 and shade in the circle. Then shade in the portion of the number line to the left of 4, that is, the set of numbers less than 4.

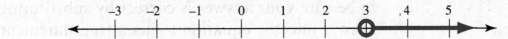

B. Graph the solution set of $y > 3$.

We are looking for the set of numbers greater than 3 but not including 3. Draw a circle around 3 but do not shade it; then highlight the numbers to the right.

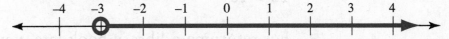

C. Graph the solution set to $-3g + 1 \leq 10$.

First, simplify the inequality.

$$-3g + 1 \leq 10$$
$$\underline{\ -1 = -1}$$
$$-3g \quad \leq 9$$

$$\frac{-3g}{-3} \quad \geq \frac{9}{-3}$$

$$g \quad \geq -3$$

Notice that the sign of $\leq$ was changed to $\geq$ when dividing by -3. The sign will *always* change when dividing by a negative number.

The graph of the correct solution is below.

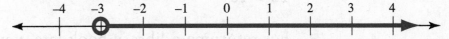

Suppose you did not change the sign. Your inequality would be $x \leq -3$. Try substituting -4 for g in the original inequality and see what happens.

$$-3g + 1 \leq 10$$
$$(-3)(-4) + 1 \leq 10$$
$$12 + 1 \leq 10$$

You get $13 \leq 10$, which is false. So remember to change the sign of the inequality when dividing by a negative number.

WRITING EQUATIONS AND INEQUALITIES FROM WORD PROBLEMS

Equations help us to solve problems more quickly. Read and understand what the problem is asking of you. Your problem-solving skills will help you to determine which information is needed. Use keys words from the problem to write the equation and then solve it. Check your work to be sure your answer is correct by substituting your answer into the equation to get a true statement.

EXAMPLES

A. Jorge paid $18 for baseballs priced at $1.50 each. How many baseballs did he buy?

Let b represent the number of baseballs purchased.
(Number of baseballs) (cost of each) = total paid
$b \cdot 1.50 = 18$
$1.50b = 18$
$b = \dfrac{18}{1.5}$
$b = 12$ Jorge purchased 12 baseballs.

Check your answer:
Does 12 • $1.50 = $18?
$12 \cdot 1.5 = 18$
$18 = 18$

B. For returning a library book late, you are charged $6.00, which includes $2.00 plus a fine of $0.50 per day. How many days did you keep the book beyond its due date?

$2 + (0.50)(d) = \$6$
$2 + 0.5d = 6$
$2 + (-2) + 0.5d = 6 + (-2)$
$0.5d = 4$
$0.5d \cdot 10 = 4 \cdot 10$ (move the decimal point one place to the right)

$5d = 40$

$d = 8$

The book was 8 days late.

Check:

Does $2 + (0.50)(8) = \$6$?

$2 + 4 = 6$

$6 = 6$

C. The eighth-grade class is filling small bags with cookies to sell at a fund-raiser. Each bag will contain 5 cookies. How many bags can they fill using 14 dozen cookies?

Let b represent the number of bags. Five cookies go in each bag. The class has $14 \cdot 12$, or 168, individual cookies to package.

$5b = 168$

$b = \dfrac{168}{5} = 33.6$

The class can fill 33 bags.

Check:

33 bags with 5 cookies in each will use $33 \cdot 5$ or 165 cookies with 3 cookies left, which is not enough for another bag. The class needs 170 cookies to fill 34 bags, but they have only 168 cookies.

D. The sum of three consecutive odd integers is 45. Find the integers.

Let x be the first integer, $x + 2$ the second, and $x + 4$ the third.

$x + x + 2 + x + 4 = 45$

$3x + 6 = 45$

$3x + 6 - 6 = 45 - 6$

$3x = 39$

$x = 13$

Therefore, the integers are 13, 15, and 17.

Check:

All three integers are odd. $13 + 15 + 17 = 45$

E. The school bookstore advertised folders at 75 cents each and pens at 50 cents each. The folder sales were $43.50. A total of $85.00 was collected. How many pens were sold?

Let p represent the number of pens sold. The income from the folders is given at $43.50. The income from the pens is the number of pens sold multiplied by the cost of each, 50 cents.

Folder sales + pen sales = $85

$$43.50 + 0.50p = 85$$
$$43.50 + (-43.5) + 0.50p = 85 + (-43.5)$$
$$0.5p = 41.5$$
$$\frac{0.5p}{0.5} = \frac{41.5}{0.5}$$
$$p = 83$$

Thus, 83 pens were sold.

Check:
Amount for folders + amount for pens = $85
$$\$43.50 + (83)(0.50) = \$85$$
$$\$43.50 + \$41.50 = \$85$$
$$\$85 = \$85$$

F. Jack wants to buy a bike for $150.00. So far, he has $63 saved. To earn money, he mows his neighbors' lawns at a charge of $20 per house. Find the least number of lawns he has to mow to pay for his bike.

Let x represent the number of lawns Jack has to mow. From the problem, you know that he needs a minimum of $150 and has only $63. Now write the inequality and solve.

$$20x + 63 \geq 150$$
$$\underline{ -63 = -63}$$
$$20x \geq 87$$
$$x \geq \frac{87}{20} \text{ or } 4\frac{7}{20}$$

Thus, Jack must mow at least 5 more lawns to get his bike.

G. Ted's boat can hold a maximum of 1,800 pounds of weight, including gas, equipment, and people. The total weight of the equipment and gas is 960 pounds. How many people can Ted take on the boat if the average weight of his friends is 200 pounds?

Since the weight cannot exceed 1,800 pounds, use the inequality symbol $\leq$.

$$960 + 200p \leq 1800$$

$$\frac{-960 \qquad\quad = -960}{200p \leq 840}$$

$$p \leq \frac{840}{200} \text{ or } 4\frac{1}{5}$$

So 4 of Ted's friends can go on his boat.

TEST YOUR SKILLS   (See page 160 for answers.)

1. Which expression represents *five less than twelve times a number?*
   (A) $5 - 12n$ (C) $12(5) - n$
   (B) $12 - 5n$ (D) $12n - 5$

2. The English phrase that best describes $\dfrac{8d + 5}{3}$ is ___.

   (A) The sum of eight times a number and five then divided by three
   (B) Eight times a number divided by three plus five
   (C) Five is added to eight times a number divided by three
   (D) Eight times a number is added to five divided by three

3. Which number is *not* in the solution set to $x + 2 \le -1$?
   (A) 0   (B) −3   (C) −5   (D) −7

4. Evaluate $\dfrac{6 - w^2}{5}$ when $w = -9$.

5. Mr. Wilson rented a truck to move some furniture. The company charged $50 for the first hour and $10 for each additional hour after the first.
   A. Write an equation to find the cost, c, of the rental.
   B. Use your equation to find the cost if Mr. Wilson returned the truck after 4 hours.

6. A company claims that the life of one of their lightbulbs is 2,000 hours. If the light is on 6 hours per day, will the bulb last more than a year? Explain how you arrived at your answer.

7. An elevator's weight limit is 1,000 pounds.
    A. Write an inequality to find the number of 90-pound students, s, who can use the elevator at the same time.
    B. Solve your inequality to find s.

8. Use the function $y = 0.5x$ to complete a table of values where the values of x are –2, –1, 0, 1, and 2.

9. For delivering flyers that advertise a new coffee shop in your neighborhood, you were paid $3 an hour plus 2 cents per flyer. You delivered 220 flyers and were paid a total of $22.40. Write an equation to find the number of hours, h, you worked and then solve the equation.

10. Sarah entered an equation into her calculator. Then she gave the calculator to you to enter a number. The table below shows the numbers you entered and the calculator results. What equation did Sarah enter into her calculator?

Input (x)	2	–3	5	0	–1	4	–2
Output (y)	9	–11	21	1	–3	17	–7

11. If a ream of copy paper weighs 4 pounds and there are 10 reams in a box, how many boxes can be placed on a cart that has a weight limit of 700 pounds? Explain how you arrived at your answer.

12. The cost of a first-class letter is $0.44 for the first ounce and $0.17 for each ounce after that.
    A. Write an equation to find the cost of mailing a letter.
    B. Use that equation to find the cost of mailing a letter that weighs 7 ounces. Show your work.

13. Graph the solution to the inequality $6y + 5 \geq 2y + 7$.

14. Write the inequality shown on the graph below.

15. Marie would like to purchase socks that cost $1.98 per pair. She has $25.00 to spend.
   A. Write an inequality that can be used to find the number of pairs she can buy.
   B. Use your inequality to solve this problem.

Chapter 4

Ratio, Proportion, and Percent

RATIO AND PROPORTION

A ratio is used to compare two quantities. If 12 girls and 16 boys are in your class, the ratio of girls to boys is written as 12 to 16, 12:16, or $\frac{12}{16}$, which can be reduced to $\frac{3}{4}$. This means that for every 3 girls in your class, there are 4 boys.

Two ratios connected with an equal sign form a *proportion*. Proportions are used to solve many problems, including scale drawings. If you know three of the four numbers in a proportion, you can easily solve to find the missing number.

These two circles are congruent but divided differently.

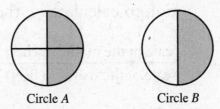

Circle *A*   Circle *B*

Circle *A* has $\frac{2}{4}$ shaded, and circle *B* has $\frac{1}{2}$ shaded. Both circles have the same area shaded.

$$\frac{2}{4} \bowtie \frac{1}{2}$$

The *cross products* in a proportion are always equal. In this example, 2 • 2 = 4 • 1. This fact about cross products is used to find the missing number in a proportion.

EXAMPLES

A. An architect's plans are on a scale of 1:3, or $\frac{1}{3}$, which means that 1 inch on the plans equals 3 feet in real life. How wide is the garage if the plans measure 6 inches?

$$\frac{1 \text{ inch}}{3 \text{ feet}} = \frac{6 \text{ inches}}{x \text{ feet}}$$

In this proportion, both numerators represent the scale from the plans, while both denominators represent the actual size. Solve this proportion using cross products.

$1 • x = 6 • 3$
$x = 18$

If 1 inch represents 3 feet, then 6 inches represent 18 feet. The garage is 18 feet wide.

B. The ship *Titanic* was 882 feet long. A scale model of the *Titanic* is $\frac{1}{63}$ of the actual size. Set up a proportion to calculate m, the length of the model in feet.

$$\frac{\text{scale of the model (inches)}}{\text{scale of the real ship (feet)}} = \frac{\text{total length of the model (inches)}}{\text{total length of the real ship (feet)}}$$

Both numerators refer to the model, while both denominators refer to the actual size of the ship. The proportion is $\frac{1}{63} = \frac{m}{882}$.

C. On a map, the distance between two cities measures 3 inches. The scale of the map is 30 miles = 0.25 inch. How far apart are the cities?

   Set up a proportion with miles in both numerators and inches in both denominators.

$$\frac{30 \text{ miles}}{0.25 \text{ inch}} = \frac{x \text{ miles}}{3 \text{ inches}}$$

$$30(3) = 0.25x$$

$$90 = 0.25x$$

$$\frac{90}{0.25} = \frac{0.25x}{0.25}$$

$$360 = x$$

Thus, the cities are 360 miles apart.

D. One dozen oranges costs $2.49. You paid $13.25 for six dozen oranges. Were you charged a lower price per dozen because of the larger quantity?

   Set up a proportion and see if the cross products are equal.

$$\frac{1 \text{ dozen}}{2.49} \stackrel{?}{=} \frac{6 \text{ dozen}}{13.25}$$

$$13.25 \stackrel{?}{=} 2.49(6)$$

$$13.25 \neq 14.94$$

At $2.49 a dozen, six dozen oranges should cost $14.94. So, yes, you were given a discount.

E. A recipe for punch calls for two 2-liter bottles of ginger ale for each pint of sherbet. The refreshment committee has 12 bottles of ginger ale. How many pints of sherbet are needed?

$$\frac{2 \text{ bottles}}{1 \text{ pint}} = \frac{12 \text{ bottles}}{x \text{ pints}}$$

$$2x = 12$$

$$x = 6$$

The committee needs to purchase 6 pints of sherbet.

F. In the example above, can you set up a different proportion to solve this problem?

The above proportion compares bottles to pints. Another proportion can be set up so the amount in the recipe can be compared to the amount needed.

$$\frac{2 \text{ bottles}}{12 \text{ bottles}} = \frac{1 \text{ pint}}{x \text{ pints}}$$

Cross multiplying gives you $2x = 12$ or 6 pints of sherbet. You cannot mix the quantities. The proportion below is not set up correctly.

$$\frac{2 \text{ bottles}}{1 \text{ pint}} = \frac{x \text{ pints}}{12 \text{ bottles}}$$

G. Which is larger, $\frac{3}{4}$ or $\frac{2}{5}$?

Set up a proportion and check.

$$\frac{3}{4} \; ? \; \frac{2}{5}$$

$$3(5) \; ? \; 2(4)$$

$$15 > 8$$

$$\frac{3}{4} > \frac{2}{5}$$

What you really did was multiply both ratios by the least common multiple of 4 and 5, which is 20.

$$\frac{3}{4}\left(\frac{20}{1}\right) \;?\; \frac{2}{5}\left(\frac{20}{1}\right)$$

$$\frac{60}{4} \;?\; \frac{40}{5}$$

$$15 > 8$$

When you multiply the factors, write the product under the fraction with the numerator you used. The product of 15 is written under the fraction $\frac{3}{4}$.

H. A 28-ounce bottle of ketchup is priced at $1.19, and another brand costs $1.89 for a 32-ounce bottle. Which is the better offer?

Set up a proportion and compare the cross products.

$$\frac{\$1.19}{28 \text{ ounces}} = \frac{\$1.89}{32 \text{ ounces}}$$

$$1.19(32) = 1.89(28)$$

$$38.08 \neq 52.92$$

$38.08 < 52.92$ means the $1.19 bottle is less expensive.

CHANGING FRACTIONS TO DECIMALS

A fraction is another name for a rational number, which is a number written in the form $\frac{a}{b}$ where a and b are integers and $b \neq 0$. In a *proper fraction*, the numerator (a) is less than the value of the denominator (b), such as $\frac{3}{5}$ or $\frac{-4}{15}$. The numerator is larger than the denominator in an *improper fraction*, such as $\frac{28}{11}$ or $\frac{-32}{21}$.

To change a fraction into a decimal number, divide the numerator by the denominator. The digits in a rational decimal number can terminate (0.25) or repeat in a pattern (0.284284284... *or* $0.\overline{284}$).

EXAMPLES

Find the decimal equivalent for the following fractions.

A. $\dfrac{3}{5} = 3 \div 5 = 0.6$

B. $\dfrac{17}{20} = 17 \div 20 = 0.85$

C. $\dfrac{7}{9} = 0.7777... = 0.\overline{7}$

D. $\dfrac{2}{11} = 0.18181818... = 0.\overline{18}$

When you notice a repeating pattern, draw a line over the digit or group of digits that repeat. Calculators round or truncate. Suppose your calculator has an eight-digit display. The answer may look like *0.7777777* when dividing 7 by 9 if your calculator truncates (drops the rest of the digits). The answer may look like *0.7777778* if your calculator rounds.

CHANGING DECIMALS TO PERCENTS

To change a decimal to a percent, move the decimal point two places to the right and add a percent sign.

EXAMPLES

A. 0.47 becomes 47%.

B. 2.835 becomes 283.5%.

C. 0.009 becomes 0.9%.

CHANGING FRACTIONS TO PERCENTS

Percents are based on *parts per hundred*. To change a fraction to a percent, find an equivalent fraction that has a denominator of 100. Write the numerator with a percent sign.

EXAMPLES

A. $\dfrac{4}{100}$ becomes 4%.

B. $\dfrac{18}{25}$ becomes $\dfrac{18 \times 4}{25 \times 4} = \dfrac{72}{100} = 72\%$.

C. $\dfrac{3}{8} = \dfrac{375}{1,000} = \dfrac{375 \div 10}{1,000 \div 10} = \dfrac{37.5}{100} = 37.5\%$

D. $\dfrac{14}{10,000} = \dfrac{14 \div 100}{10,000 \div 100} = \dfrac{0.14}{100} = 0.14\%$

E. To change $\dfrac{2}{3}$ to a percent, divide 2 by 3 and then change the decimal to a percent.

$\dfrac{2}{3} = 2 \div 3 = 0.6666\ldots = 66.\overline{6}\%$

SOLVING PERCENT PROBLEMS

Proportions can be used to solve percent problems using the proportion

$$\frac{\text{is}}{\text{of}} = \frac{\text{rate}}{100}$$

Read the problem carefully and substitute the correct numbers into the proportion. Then solve for the missing number.

EXAMPLES

A. Of his $20 allowance, Mark saved $7. What percent did he save?

The amount Mark saved is $7. Replace *is* with 7 and replace *of* with 20 in the proportion. Let x be the rate.

$$\frac{7}{20} = \frac{x}{100}$$
$$700 = 20x$$
$$\frac{700}{20} = \frac{20x}{20}$$
$$35 = x$$

Mark saved 35% of his $20 allowance.

B. During a 30%-off sale, the price of a DVD player is now $105. Find the original price of the DVD player.

The price of $105 is the amount we paid. In other words, $105 is equal to 70% of the original price.

$$\frac{105}{x} = \frac{70}{100}$$
$$10,500 = 70x$$
$$150 = x$$

The original price was $150.

C. The Happy Feet Shoe Company can close when a minimum of 20% of the employees call in sick. If the factory has 690 employees, how many workers need to be absent before the company can close?

$$\frac{x}{690} = \frac{20}{100}$$
$$100x = 690(20)$$
$$100x = 13,800$$
$$x = 138$$

The factory can close when 138 people call in sick.

D. How much of a gratuity should you leave for your server if the restaurant bill is $37.50 and you want to leave a 20% tip?

You might want to substitute 37.50 for *is*, but you are calculating the tip. The problem can be rephrased as *what is 20% of $37.50?*

$$\frac{x}{37.50} = \frac{20}{100}$$
$$100x = 20(37.50)$$
$$100x = 750$$
$$x = 7.5$$

You should leave $7.50 for your server.

E. The gas station listed regular gas at $2.50 per gallon yesterday. Today, the price is $2.55. By what percent did the price increase?

The price increased by $0.05. Find the percent of this increase using a proportion.

$$\frac{\text{Difference in price}}{\text{Original price}} = \frac{x}{100} \Rightarrow \frac{0.05}{2.50} = \frac{x}{100}$$
$$0.05(100) = 2.5x$$
$$5 = 2.5x$$
$$2 = x$$

So, the gas price increased by 2%.

F. The admissions department at a local college is allowed to accept a certain percentage of students from those who apply. How many students can be accepted if 18,500 applied and the department can accept 3%?

The problem can be rewritten as *3% of 18,500 is the number of accepted students*. Write an equation by translating the English sentence into algebra and solve it. Change the percent to its decimal equivalent. The word *of* means multiply.

$$\frac{x}{18,500} = \frac{3}{100}$$
$$100x = 3(18,500)$$
$$100x = 55,500$$
$$x = 555$$

The college can accept 555 students.

G. Calculate the total bill of a new car priced at $19,850 with a state tax rate of $8\frac{3}{4}\%$.

   This can be a one-step or a two-step problem. A two-step problem calculates the tax in the first step and adds the tax to the car's price in the second step.
   1st step: $19,850(0.0875) \Rightarrow 1,736.88$ tax (rounded to the nearest hundredth)
   2nd step: $19,850 + \$1,736.88 = \$21,586.88$ total cost of car

   In the second step, we added 100% of the car plus $8\frac{3}{4}\%$ of the price of the car in tax. Therefore,

   $108\frac{3}{4}\%$ of the car's price = total cost of car

   $1.0875(\$19,850) = \$21,586.88$ (rounded)
   Either way you solve, you get the same answer.

H. Nicole receives a commission of 4% on the base price for each car she sells. How much did she receive from a sale if the base price was $23,400?

   4% of $23,400 = (0.04)(23400) = \$936$
   Nicole received $936 as a commission.

I. The sales price of a leather jacket was $195. It was regularly priced at $300. What percent was saved? $300 – $195, or $105, was saved. Solve using a proportion.

$$\frac{105}{300} = \frac{x}{100}$$
$$105(100) = 300x$$
$$10,500 = 300x$$
$$35 = x$$

35% was saved.

J. This year, heating oil customers are paying 120% of the price they paid last year. What are they paying now if they paid $140 last year?

140(1.2) = 168

Customers are paying $168 if they paid $140 last year.

K. If the amount of increase in the price of heating oil is 120%, calculate the cost to customers if they paid $140 last year.
   Since the amount of increase is 120%, customers are paying 100% + 120%, or 220%, of last year's price. 2.2(140) = 308
Customers now pay $308.

L. There were 11,374 students who attended the championship game. Of these, 1,683 students painted their faces with the school colors. To the nearest whole number, approximate the percentage that painted their faces.

$$\frac{1,683}{11,374} \approx 0.1479 \approx 0.15 = 15\%$$

About 15% of the spectators painted their faces.

M. After assessing the damage done by bunnies over the winter, the gardeners at the Willowcreek Botanical Gardens had to replace tulip bulbs. Last spring, there were 28,693 bulbs. This spring, only 25,612 sprouted. What percentage of bulbs had to be replaced so there was the same number of bulbs as last spring?

28,693 – 25,612 = 3,081 needed to be replaced

$$\frac{3,081}{28,693} \approx 0.1073 \approx 0.11 = 11\%$$

Approximately 11% of the bulbs had to be replaced.

INTEREST

To calculate simple interest on an investment, the formula $i = prt$ is used. The amount of interest, i, is the product of the *p*rincipal (the amount invested), the *r*ate, and the amount of *t*ime in years.

EXAMPLES

A. You deposited $250 in the bank at an annual percentage rate of 3.5%. How much interest did you receive at the end of 2 years?

$i = prt$
$i = (250)(0.035)(2)$
$i = 17.5$
You received $17.50 in interest.

B. After 18 months, your uncle's deposit of $1,500 made $56.25 in interest. Find the interest rate.

$56.25 = (1,500)(1.5)r$
$56.25 = 2,250r$

$$\frac{56.25}{2,250} = r$$

$0.025 = r$
$r = 2.5\%$
The interest rate was 2.5%.

C. Sam's investment earned $630 at a rate of 2.25% at the end of 1 year. What amount did he invest?

$$i = prt$$
$$630 = p(0.0225)(1)$$
$$630 = 0.0225p$$
$$\frac{630}{0.0225} = \frac{0.0225p}{0.0225}$$
$$28,000 = p$$

Sam invested $28,000.

TEST YOUR SKILLS   (See page 163 for answers.)

1. When written as a percent, $\frac{2}{5}$ is equal to which choice?
(A) 2.5%   (B) 40%   (C) 10%   (D) 4%

2. A coin set is appraised at $10,000. If the value increases by 3%, how much more will the set be worth at the end of this year?
(A) $3   (B) $30   (C) $300   (D) $3,000

3. What is 0.6% of 1,200?
(A) 200   (B) 72   (C) 7.2   (D) 720

4. After 1 year, Micky's statement showed a balance of $15,080. If she earned 4% interest, find the amount she originally invested.

5. How many more squares should be shaded so that 28% are left white? Explain your answer.

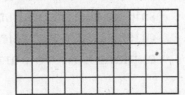

6. At a pizza party, Shelly ate $\frac{3}{10}$ of the pizza and Kelly ate 10%. How much was left for the other guests? Show your work.

7. The directions on a container of iced tea mix state that you should use 1 cup to make 1 quart. One quart is equal to 32 ounces. The only container you have holds 24 ounces. How much mix should you use to make 24 ounces of iced tea? Explain how you arrived at your answer.

8. The 8% sales tax was increased by $\frac{1}{4}$%.
   A. Write the new tax rate as a decimal number.
   B. Write an equation to find the sales tax at this new rate, t, to be paid on four bookcases priced at x dollars each.

9. The Smithfield Middle School has an enrollment of 700 students. One day during the flu season, 95 students were absent. The school can close because of illness if 15% of the students are absent. Could Smithfield close that day? Explain your answer.

10. A jeweler sold a necklace for 250% of the amount he paid for it. If the necklace sold for $2,085, how much did it cost him?

11. Howie went shopping for a new calculator. The first store's price was $35.95. There was a 10%-off sale. The sales tax was 8%. The second store had the same calculator for $39.95 at 20% off. There was a *no tax* sale at the second store. Which store had the better offer? Use your calculator to solve this problem and explain how you arrived at your answer.

12. At the Greenfield School, only 2.4% of the students, or 21, are left-handed. How many students are in the school?

13. Last year's lawn-mowing services cost $70 per month. This year's monthly cost is $101.50. What is the percentage of increase?

14. Find the interest rate if after 6 months, $800 earned $4.80 in interest.

15. Ms. Harper was tracking the number of students who solved the extra-credit question on her math tests. Since very few students attempted the question, she decided that if less than 5% tried the problem, she would not include extra-credit problems after this test. Only 7 of the 173 students attempted the problem. What was her decision?

Chapter 5

Measurement

CUSTOMARY SYSTEM

Before there were instruments of measurement such as rulers or metersticks, people used their body parts. For example, a yard was the distance between a person's nose and the end of his hand, or a foot was the length of the measurer's foot. Because not all people are created equal in size, a standardized system was developed. We in the United States use the Customary System most of the time but also use the Metric System.

Customary System

Length	Weight	Capacity
12 inches (in.) = 1 foot (ft)	16 ounces (oz) = 1 pound (lb)	8 ounces = 1 cup
3 feet = 1 yard (yd)	2,000 pounds = 1 ton	2 cups = 1 pint (pt)
5,280 feet = 1 mile (mi)		2 pints = 1 quart (qt)
		4 quarts = 1 gallon (gal)

EXAMPLES

A. How many yards are in a mile?

$$\frac{5,280\,\dfrac{\text{ft}}{\text{mi}}}{3\,\dfrac{\text{ft}}{\text{yd}}} = 1,760\,\frac{\text{yd}}{\text{mi}}$$

There are 1,760 yards in a mile.

B. Can 6 cups of orange juice fit in a 1-quart container?

If 2 cups = 1 pint and 2 pints = 1 quart, then 4 cups = 1 quart. No, 6 cups will not fit in a 1-quart (or 4-cup) container.

C. Which is larger: $\frac{1}{4}$ ton or 600 pounds?

Six hundred pounds is greater because $\frac{1}{4}$ ton is 500 pounds.

D. What is the length of the line below in inches?

_____

The length of the line is 5 inches.

METRIC SYSTEM

The most widely used system of measurement in the world is the Metric System. It is based on powers of 10. The basic units of measure are *meter* (length), *gram* (weight), and *liter* (capacity).

Metric System

Unit	1/1,000	1/100	1/10	1	10	100	1,000
Length	millimeter	centimeter	decimeter	meter	dekameter	hectometer	kilometer
Weight	milligram	centigram	decigram	gram	dekagram	hectogram	kilogram
Capacity	milliliter	centiliter	deciliter	liter	dekaliter	hectoliter	kiloliter

The prefixes (*milli* $\Rightarrow \dfrac{1}{1,000}$, *centi* $\Rightarrow \dfrac{1}{100}$, *deci* $\Rightarrow \dfrac{1}{10}$,

deka $\Rightarrow$ 10, *hecto* $\Rightarrow$ 100, and *kilo* $\Rightarrow$ 1,000) added to the basic units give you measures in the Metric System. The most commonly used measures are shown in bold in the table. A paperclip weighs about 1 gram. A doorknob is about 1 meter from the floor. A liter is a little larger than 1 quart. Because this system is based on powers of 10, it is easier to use and to convert measures than in the Customary System.

EXAMPLES

A. How many milliliters are in 80 liters?

In the Metric System table, find *milliliter* and move to *liter*. You moved three places to the right. So move the decimal point in 80 three places to the right. There are 80,000 milliliters in 80 liters.

B. If one paperclip weighs 1 gram and the empty box itself weighs 3 grams, find the weight in grams of 6 boxes of paperclips that each contain 50 paperclips.

There are 6 boxes at 53 grams each (3 grams for the box and 50 grams for the paperclips). The total weight is 53(6) or 318 grams.

C. How many meters are in 735 centimeters?

Find *meter* in the Metric System table and move to *centimeter*. You moved two places to the left. So move the decimal point in 735 two places to the left. So 735 centimeters equals 7.35 meters.

D. Find the length of the line below in both centimeters and millimeters.

_____

The line measures 7 centimeters or 70 millimeters.

CONVERTING BETWEEN SYSTEMS

Metric to U.S.	U.S. to Metric
1 km ≈ $\frac{5}{8}$ mi	1 mi ≈ 1.6 km
1 kg ≈ 2.2 lb	1 lb ≈ 0.45 kg
1 L ≈ 1.06 qt	1 qt ≈ 0.95 L

The Metric System is used in Canada. Car speedometers show speeds in both systems. If you see a speed limit sign posted at 80 kilometers per hour, you are traveling at approximately 50 miles per hour. Prices at Canadian gas stations are shown as the cost per liter.

Suppose the cost per gallon of gas in Buffalo is $3.73 and the cost per Canadian liter is $1.08. Is the gas cheaper in Canada or in Buffalo?

If 1 quart ≈ 0.95 liter, then 4 quarts ≈ 4(0.95) or 3.8 liters.

$$\$3.73 \div 3.8 \approx \$0.98 \text{ per liter in Buffalo}$$

You can also solve this problem by multiplying the cost per liter by 3.8 to get the cost of a gallon of gas in Canadian funds.

$$\$1.08 \times 3.8 \approx \$4.10 \text{ per gallon in Canada}$$

Either way, the cost of gas is lower in Buffalo.

TEMPERATURE CONVERSION

Water freezes at 0°C or 32°F. It boils at 100°C or 212°F. Room temperature is about 70°F or 21°C. The formulas used to convert the temperature between Fahrenheit and Celsius are

$$F = \frac{9}{5}C + 32 \qquad \text{and} \qquad C = \frac{5}{9}(F - 32)$$

EXAMPLES

A. Change 50°F to the nearest whole degree in Celsius.

$$C = \frac{5}{9}(F - 32)$$

$$C = \frac{5}{9}(50 - 32)$$

$$C = \frac{5}{9}(18)$$

$$C = 10$$

So 50°F is about 10°C.

B. A thermometer inside a refrigerator shows the temperature as –8°C. Find the temperature in Fahrenheit to the nearest degree.

$$F = \frac{9}{5}C + 32$$

$$F = \frac{9}{5}(-8) + 32$$

$$F = \frac{-72}{5} + 32$$

$$F(5) = \left(\frac{-72}{5} + 32\right)(5)$$

$$5F = -72 + 160$$

$$5F = 88$$

$$F \approx 17.6$$

To the nearest degree, –8°C is about 18°F.

MONETARY CONVERSION

Vacationers and business travelers need to understand how to convert between their system of currency and the system of the country they are visiting to be sure they are getting the correct amount in return. In the examples that follow, the exchange rate is shown for each situation. The actual exchange rates change daily.

EXAMPLES

A. Suppose 1 euro (€) is equivalent to $1.20 USD (U.S. dollars). Find the value of $1 in euros.

$$\frac{1}{1.20} \approx 0.83$$

$$€1 \approx \$0.83$$

B. If you see a souvenir shirt priced at €27 and you have $35, do you have enough money to buy the shirt?
  Use a proportion to find the equivalent dollar amount of €27 by comparing $\dfrac{1 \text{ euro}}{\$1.20} = \dfrac{\text{euros}}{\text{dollars}}$.

$$\frac{1}{1.2} = \frac{27}{x}$$

$$x = 27(1.20) = \$32.40$$

Yes, you have enough to buy the shirt.

C. Calculate the exchange of $20 into euros.

$$\frac{\$1}{0.83 \text{ euro}} = \frac{\$20}{x \text{ euros}}$$

$$x = (20)(0.83)$$

$$x = €16.60$$

You can also solve this problem by multiplying 20(0.83) to get €16.60.

Multiply dollars by 0.83 to get euros, and multiply euros by 1.2 to get dollars at this rate.

D. Calculate the exchange of $250 USD into yen if 1 yen is $0.008.

First, find the number of yen in $1.

$$\frac{1}{0.008} = 125$$

If 125 yen are in $1, then 125(250) or 31,250 yen are in $250.

TEST YOUR SKILLS   (See page 166 for answers.)

1. How many inches equal $\frac{1}{2}$ yard?

(A) 24   (B) 12   (C) 18   (D) 6

2. 67 millimeters = ___ meters.

(A) 670   (B) 6.7   (C) 0.67   (D) 0.067

3. How many 8-ounce cups of iced tea are in 6 gallons?

(A) 12   (B) 48   (C) 64   (D) 96

4. Calculate the exchange of $55 into euros if 0.83 euro equals $1.

5. Approximate the number of meters in $\frac{1}{4}$ mile.

6. If 1 gallon of water weighs approximately 8 pounds, estimate the weight of 1 cup of water.

7. Approximate the number of kilometers between two towns if you know the distance is 18 miles.

8. A Canadian radio station gave the outside temperature as 27°C. Find the Fahrenheit equivalent to the nearest degree.

9. The room parents of the eighth-grade class are planning a picnic. They will provide three 12-ounce servings of orange drink for each of the 24 students. How many gallons of orange drink should the parents bring? Show your work.

10. Matt and Paul want to sign up for a 5-kilometer race. Because of his injured knee, Paul cannot run distances longer than 3 miles. Should Paul sign up for the race? Explain your answer.

11. Megan's family is planning a camping vacation in Canada. They will fill up the gas tank in their van before they cross the border and won't fill it up again until they get back to New York State. Megan's mom knows the range of their van is 440 miles on one tank of gas. The family wants to stay at a campground that is 270 kilometers from the border. Will they make it back to New York State without running out of gas? Explain your answer.

12. A car is traveling at 60 kilometers per hour. To the nearest whole number, find the speed in miles per hour.

13. Approximate the maximum number of liters needed to fill a 20-gallon aquarium.

14. A healthy person's body temperature is 98.6°F. What is that temperature in Celsius?

15. On a trip to Belgium, you see a watch for €23. The exchange rate is $1 = €1.08. Can you buy the watch if you have $20?

Chapter 6

Monomials and Polynomials

A *monomial,* or term, can be a variable, a constant, or a product of both. The algebraic expressions $4m - 3$, $x^2 + 25$, and $12x^3 + 7y + 2$ are called *polynomials* because they contain more than one monomial. A polynomial can be called a *binomial* when it has two terms or a *trinomial* when it has three terms.

ADDING POLYNOMIALS

To add polynomials, combine like terms.

EXAMPLES

A. $(7x^2 + 3) + (11x^2 + 5x + 8) = ?$
Regroup the like terms. Add the coefficients of the like terms.
$(7x^2 + 11x^2) + 5x + (3 + 8) =$
$18x^2 + 5x + 11$

B. $2x + 4x^2 + 3 - 8x + x^2 - 7 =$
$(4x^2 + x^2) + (2x - 8x) + (3 - 7) =$
$5x^2 - 6x - 4$

SUBTRACTING POLYNOMIALS

To subtract polynomials, change from subtraction to addition and then add the opposite of every term in the second polynomial just as you do when subtracting integers.

EXAMPLES

A. $(10a + 8b) - (4a + 5b) =$
$10a + 8b - 4a - 5b =$
$10a - 4a + 8b - 5b =$
$6a + 3b$

B. $(32x^5 + 14y^3 - 2) - (13x^5 - 3y^3 - 9) =$
$32x^5 + 14y^3 - 2 - 13x^5 + 3y^3 + 9 =$
$32x^5 - 13x^5 + 14y^3 + 3y^3 - 2 + 9 =$
$19x^5 + 17y^3 + 7$

C.
$$\begin{array}{rcrcr} 15t^2 & + & 8t & & \\ -(-19t^2 & + & & 7) \end{array} \rightarrow \begin{array}{rcrcr} 15t^2 & + & 8t & & \\ +19t^2 & & & - & 7 \\ \hline 34t^2 & + & 8t & - & 7 \end{array}$$

LAWS OF EXPONENTS

Multiplication

To multiply powers of the same base, add the exponents.

$$a^3 \cdot a^2 = (a \cdot a \cdot a) \cdot (a \cdot a) = a^{3+2} = a^5$$

EXAMPLES

A. Find the product of $a^4 \cdot b^2 \cdot a^2 \cdot b^5$.
$a^4 \cdot a^2 \cdot b^2 \cdot b^5 = a^{4+2} \cdot b^{2+5} = a^6 b^7$

B. Simplify $(5m^2)(3m^4)$.
$(5m^2)(3m^4) =$
$(5)(3) (m^2)(m^4) = 15m^{2+4} = 15m^6$

C. Multiply $(4k^5)(6p^3)(2k^3)(3p^6)$.
$(4k^5)(6p^3)(2k^3)(3p^6) =$
$(4)(6)(2)(3)(k^5)(k^3)(p^3)(p^6) = 144k^8p^9$

D. Simplify $d^2(5g^3)(4d) + 4c^2(3g)(5c^3)$.
$(5)(4)(d^2)(d)(g^3) + (4)(3)(5)(c^2)(c^3)(g) =$
$20d^3g^3 + 60c^5g$

Raising a Power to a Power

To raise a power to a power, multiply the exponents.

$$(r^5)^2 = r^5 \cdot r^5 = r^{10}$$

EXAMPLES

Simplify the following.

A. $(k^3)^5 = (k^3)(k^3)(k^3)(k^3)(k^3) = k^{15}$

B. $(4c^3)^2 = (4c^3)(4c^3) = (4)^2(c^3)^2 = 16c^6$
When the coefficient is greater than 1, distribute the exponent to both factors in the parentheses.

C. $(8s^3)^4 = (8)^4(s^3)^4 = 4,096s^{12}$

Division

To divide powers with the same base, subtract the exponents.

$$\frac{w^5}{w^3} = \frac{w \cdot w \cdot w \cdot w \cdot w}{w \cdot w \cdot w} = w^{5-3} = w^2$$

EXAMPLES

A. Divide $\frac{s^9}{s^4}$.

$\frac{s^9}{s^4} = s^{9-4} = s^5$

B. Divide $\dfrac{w^4}{w^7}$.

$$\dfrac{w^4}{w^7} = w^{4-7} = w^{-3} = \dfrac{1}{w^3}$$

When a term has a coefficient of 1 and a variable with a negative exponent, it can be written as a rational number with a numerator of 1. The denominator will be the term without the negative sign on the exponent.

C. Divide $\dfrac{24c^5}{3c^2}$.

Divide the coefficient 24 by 3, as you normally do. Then simplify the variable terms.

$$\dfrac{24c^5}{3c^2} = 8c^3$$

D. Divide $\dfrac{15h^2}{5h^7}$.

$$\dfrac{15h^2}{5h^7} = 3h^{-5} = \dfrac{3}{h^5}$$

Here the coefficient is 3, and it stays in the numerator. Substitute a number for a variable to see if the equation is true. Substitute 2 for h in the problem above as a check.

$$\dfrac{15h^2}{5h^7} = \dfrac{15(2^2)}{5(2^7)} = \dfrac{15(4)}{5(128)} = \dfrac{60}{640} = \dfrac{60 \div 20}{640 \div 20} = \dfrac{3}{32} \quad \text{and} \quad \dfrac{3}{h^5} = \dfrac{3}{2^5} = \dfrac{3}{32}$$

$$\dfrac{3}{32} = \dfrac{3}{32}$$

MULTIPLYING A MONOMIAL BY A MONOMIAL

To multiply monomials, group powers of the same base together. The coefficients are multiplied, and the exponents are added.

EXAMPLES

A. $(6b^4)(5b^2) = 6 \cdot b^4 \cdot 5 \cdot b^2 = 6 \cdot 5 \cdot b^4 \cdot b^2 = 30b^6$

B. $3xy(9x^3y^2) = 3 \cdot 9 \cdot x \cdot x^3 \cdot y \cdot y^2 = 27x^4y^3$
If you don't see an exponent on a variable, it is understood to be 1 and should be included in the sum when adding the exponents of like variables.

C. $(14d^5)(d^4k^4)(5m^3k^2) = 14 \cdot 1 \cdot 5 \cdot d^5 \cdot d^4 \cdot k^4 \cdot k^2 \cdot m^3$
$$= 70d^9k^6m^3$$

D. Evaluate $10x^3$ when $x = 2$.
$10x^3 = 10(2^3) = 10(8) = 80$
According to the order of operations, exponents are calculated before multiplying. So calculate $2^3 = 8$ and then multiply 8 by 10.

MULTIPLYING A BINOMIAL BY A MONOMIAL

The same problem can be solved in several ways. The example that follows uses algebra tiles and the distributive property to solve the same problem. Use the way you understand best.

To help understand multiplication of polynomials, algebra tiles are often used to show a geometric solution. The tiles come in three shapes: large squares, small squares, and rectangles. Each shape comes in two colors to show positive and negative quantities.

Instead of using tiles, the same procedure can be accomplished by drawing a grid. In the examples that follow, a shaded area represents a negative value. When you have the same size shape in each color, the sum of the two areas is zero. Just cross out one shape of each color.

Let's look at the problem $3y(6y + 5)$. First, draw a horizontal line and a vertical line that meet. Label the sides with the factors. In this example, the horizontal line represents $6y + 5$. Mark off a small section of the horizontal to depict the value y. Continue until there are 6 identical units of y. Do the same with a smaller unit to show 5 identical units of 1.

The vertical line represents $3y$. With the same length of y used on the horizontal line, mark off 3 identical units of y on the vertical line.

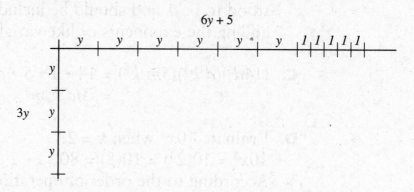

Draw the horizontal and vertical grid lines.

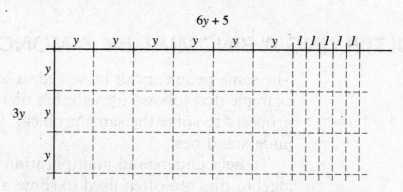

Each large square has an area of $y \cdot y$ or y^2. There are 18 large squares. Each rectangle has an area of $1 \cdot y$ or y. There are 15 rectangles.

Adding the areas gives $18y^2 + 15y$. So $3y(6y + 5)$ equals $18y^2 + 15y$.

You can also use the distributive property.

$$3y(6y + 5) = (3y \cdot 6y) + (3y \cdot 5) = 18y^2 + 15y$$

Multiply 3y by each term in the binomial. Keep the plus sign in your answer.

EXAMPLES

A. $8r(4r - 3) = (8r \cdot 4r) - (8r \cdot 3) =$

$32r^2 - 24r$

You can also change the minus sign to an addition sign and add the inverse of 3.
$8r(4r - 3) = 8r(4r + -3) =$
$(8r \cdot 4r) + (8r \cdot -3) =$
$32r^2 + -24r =$
$32r^2 - 24r$

B. $-11w^3(2w^2 - w + 7) =$
$-11w^3(2w^2 + -w + 7)$
$(-11w^3 \cdot 2w^2) + (-11w^3 \cdot -w) + (-11w^3 \cdot 7) =$
$-22w^5 + 11w^4 - 77w^3$

C. $10c^2d(3c^2 - 2cd^2 - 4d^2) = 30c^4d - 20c^3d^3 - 40c^2d^3$

SQUARING A MONOMIAL

When we square a monomial, we raise it to the power of 2. That means that we use the factors in the parentheses twice. For example, $(4d)^2 = 4d \cdot 4d = 16d^2$.

Another way to think about solving this type of problem is to raise each factor in the monomial to the power of 2.

$$(6m^2 p^4)^2 = (6)^2 \cdot (m^2)^2 \cdot (p^4)^2 = 36m^4p^8$$

MULTIPLYING A BINOMIAL BY A BINOMIAL

The product of two binomials is usually a trinomial of the form $ax^n + bx + c$, where a, b, and c are coefficients. All the problems in this chapter have $a = 1$.

Suppose a rectangle has a length of $x + 3$ and a width of $x + 2$. To find the area of this rectangle, multiply the binomials using the distributive property. A special mnemonic is used when multiplying a binomial by a binomial—*FOIL*—where *F* stands for the *first* term in each binomial, *O* for the *outside* terms, *I* for the *inside* terms, and *L* for the *last* terms.

$$(x + 3)(x + 2) =$$

$$F \rightarrow (x + 3)(x + 2) \rightarrow x \cdot x = x^2$$

$$O \rightarrow (x + 3)(x + 2) \rightarrow x \cdot +2 = +2x$$

$$I \rightarrow (x + 3)(x + 2) \rightarrow +3 \cdot x = +3x$$

$$L \rightarrow (x + 3)(x + 2) \rightarrow +3 \cdot +2 = +6$$
$$(x + 3)(x + 2) = x^2 + 2x + 3x + 6$$
$$= x^2 + 5x + 6$$

Below is the solution to the same problem using the algebra tile grid. By adding the areas together, you get $x^2 + 5x + 6$.

EXAMPLES

A. $(b - 5)(b + 2) =$

$\quad$ F $\qquad$ O $\qquad$ I $\qquad$ L

$(b \cdot b) + (b \cdot 2) + (-5 \cdot b) + (-5 \cdot 2) =$
$\quad b^2 \quad + \quad 2b \quad - \quad 5b \quad - \quad 10 \quad =$
$b^2 - 3b - 10$

B. $(3y - 1)(2y + 3)$
Construct the grid as shown in a previous example. Remember that the negative values are indicated as shaded regions.

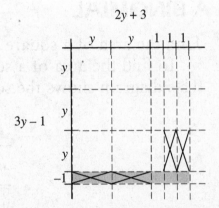

Since the rectangles represent the same area and you know that $-1 + 1 = 0$, cross out one shaded rectangle for each clear one. Then, count the areas. You are left with $6y^2 + 7y - 3$.

C. $(y + 4)(y + 4) =$
$y^2 + 4y + 4y + 16 =$
$y^2 + 8y + 16$

D. $(m + 3)(m - 3)$

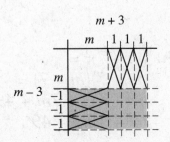

You get $m^2 - 9$. Notice that the middle term equals zero. This happens when the terms are the same but the signs are opposite in each binomial. The product, $m^2 - 9$, is called a *difference of two perfect squares*.

E. $(r + 2s)(2r + 3s) =$

$$\underbrace{}_{F} \quad \underbrace{}_{O} \quad \underbrace{}_{I} \quad \underbrace{}_{L}$$

$(r \cdot 2r) + (r \cdot 3s) + (2s \cdot 2r) + (2s \cdot 3s) =$
$\quad 2r^2 \quad + \quad 3rs \quad + \quad 4rs \quad + \quad 6s^2 \quad =$
$2r^2 + 7rs + 6s^2$

SQUARING A BINOMIAL

Find the area of a square whose side has a length of $s + 4$.

To find the area of a square, multiply the sides together. This diagram shows the solution using the algebra tile grid.

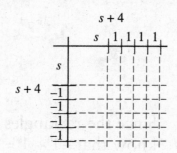

The area of the square is $s^2 + 8s + 16$.

You can also multiply $(s + 4)(s + 4)$ as you would multiply two binomials.

$$(s + 4)(s + 4) = s^2 + 4s + 4s + 16$$
$$= s^2 + 8s + 16$$

EXAMPLES

A. $(m + 7)^2 =$
$(m + 7)(m + 7) =$
$m^2 + 7m + 7m + 49 =$
$m^2 + 14m + 49$

B. $(c - 4)^2 =$
$(c - 4)(c - 4) =$
$c^2 - 4c - 4c + 16 =$
$c^2 - 8c + 16$

C. $(t + 4w)^2 =$
$(t + 4w)(t + 4w) =$
$t^2 + 4tw + 4tw + 16w^2 =$
$t^2 + 8tw + 16w^2$

D. $(g - 3)^2 =$
$(g - 3)(g - 3) =$
$g^2 - 3g - 3g + 9 =$
$g^2 - 6g + 9$

Do you see a pattern in the examples above?

$$(m + 7)^2 = m^2 + 14m + 49$$
$$(g - 3)^2 = g^2 - 6g + 9$$

Notice that $(a + b)^2 = a^2 + 2ab + b^2$. The variable a represents the first term, and b represents the second term. The middle term of the trinomial answer is the product of the two terms doubled.

EXAMPLES

A. $(r + 9)^2 =$
$(r)^2 + 2(9r) + 9^2 =$
$r^2 + 18r + 81$

B. $(p - 5)^2 =$
$(p)^2 + 2(-5p) + (-5)^2 =$
$p^2 - 10p + 25$

DIVIDING A POLYNOMIAL BY A MONOMIAL

Divide each term of the polynomial by the monomial. We know that $\dfrac{3 + 6}{14} = \dfrac{3}{14} + \dfrac{6}{14}$. Rewrite the problem so the terms in the numerator are written as individual fractions with the given denominator.

EXAMPLES

A. $\dfrac{24a^4b^2 + 16ab^3}{8ab^2} =$

$\dfrac{24a^4b^2}{8ab^2} + \dfrac{16ab^3}{8ab^2} =$

$3a^3 + 2b$

B. $\dfrac{9k^5m^3 + 12k^3m - 6km^3}{3km} =$

$\dfrac{9k^5m^3}{3km} + \dfrac{12k^3m}{3km} - \dfrac{6km^3}{3km} =$

$3k^4m^2 + 4k^2 - 2m^2$

C. $\dfrac{36x^6y^7 + 24xy^4 + 18}{6x^2y} =$

$\dfrac{36x^6y^7}{6x^2y} + \dfrac{24xy^4}{6x^2y} + \dfrac{18}{6x^2y} =$

$6x^4y^6 + \dfrac{4y^3}{x} + \dfrac{3}{x^2y}$

FACTORING ALGEBRAIC EXPRESSIONS

Factoring an expression is like using the distributive property backward. The GCF is used to find the factors of an expression. In factoring $12x^3 + 18x^2$, find the GCF of 12 and 18 and then find the GCF of x^3 and x^2.

The GCF of 12 and 18 is 6. The GCF of x^3 and x^2 is x^2. Now use $6x^2$ as the divisor of $12x^3 + 18x^2$.

$$\frac{12x^3 + 18x^2}{6x^2} =$$

$$\frac{12x^3}{6x^2} + \frac{18x^2}{6x^2} =$$

$$2x + 3$$

So $12x^3 + 18x^2 = 6x^2(2x + 3)$. To check if your result is correct, multiply using the distributive property.

EXAMPLES

A. Factor $(15k^7m^4 - 12k^4m^6)$.

The GCF(15, 12) = 3 and the GCF(k^7m^4, k^4m^6) = k^4m^4. Divide by $3k^4m^4$ to find the other factor.

$$\frac{15k^7m^4 - 12k^4m^6}{3k^4m^4} =$$

$$\frac{15k^7m^4}{3k^4m^4} - \frac{12k^4m^6}{3k^4m^4} =$$

$$5k^3 - 4m^2$$

$$15k^7m^4 - 12k^4m^6 = 3k^4m^4(5k^3 - 4m^2)$$

B. Factor $32b^6c^{11} - 24b^5c^4 + 36c^3$.

The GCF is $4c^3$. Notice that the variable b is not in the GCF because the last term does not have b as a factor.

$$\frac{32b^6c^{11} - 24b^5c^4 + 36c^3}{4c^3} =$$

$$8b^6c^8 - 6b^5c + 9$$

$$32b^6c^{11} - 24b^5c^4 + 36c^3 = 4c^3(8b^6c^8 - 6b^5c + 9)$$

FACTORING TRINOMIALS OF THE FORM $ax^2 + bx + c$

To factor the trinomial $x^2 + 7x + 10$ into two binomials, keep the FOIL method in mind. Start the process by writing two sets of parentheses for the binomial factors. The term, x^2, has two factors of x. So the x goes in the first spot in each of the binomial factors.

$$(x \quad)(x \quad)$$

Find two factors of +10 whose sum is +7. The factors are +2 and +5.

$$(x \quad 2)(x \quad 5)$$

Since the sign on the 10 is positive, we know that both of the signs in the factors will be the same. In this case, both are positive.

$$(x + 2)(x + 5)$$

To check your answer, multiply the binomials using the FOIL method.

$$(x + 2)(x + 5) =$$
$$x^2 + 5x + 2x + 10 =$$
$$x^2 + 7x + 10$$

Remember that the signs play an important role. Review the rules for operations on integers in Chapter 2.

EXAMPLES

A. Factor $m^2 - 9m + 14$.

Because 14 is positive, the signs of the binomials will be the same since the product of integers with the same sign is positive. Since 9 is negative, both signs will be negative. Write out the factors of +14 whose sum is −9. Since −1 and −14 produce −15, they are not correct.

Since –2 and –7 give you –9, they are correct. Now, fill in the parentheses and check your answer.

$(m - 2)(m - 7) =$

$m^2 - 7m - 2m + 14 =$

$m^2 - 9m + 14$

B. Factor $t^2 - 3t - 18$.

Since 18 is negative, the signs of the binomial factors will be different. Find two factors of –18 whose sum is –3. The factors –6 and +3 will work.

$(t - 6)(t + 3) =$

$t^2 + 3t - 6t - 18 =$

$t^2 - 3t - 18$

C. Factor $h^2 - 2h - 35$.

Find two factors of –35 whose sum is –2. The signs of the factors will be different.

$(h - 7)(h + 5) =$

$h^2 + 5h - 7h - 35 =$

$h^2 - 2h - 35$

D. Factor $x^2 + 10x + 16$.

Find two factors of 16 whose sum is 10. The binomial factors are $(x + 2)$ and $(x + 8)$. Multiply to check.

$(x + 2)(x + 8) =$

$x^2 + 8x + 2x + 16 =$

$x^2 + 10x + 16$

TEST YOUR SKILLS   (See page 168 for answers.)

1. When simplified, $7m + 9y + 12y - 4m =$

(A) $3m + 21y$ (C) $16my + 8ym$

(B) $-28m^2 + 108y^2$ (D) $-11m + 21y$

2. The factors of $k^2 - k - 30$ are

(A) $(k - 5)$ and $(k - 6)$ (C) $(k - 6)$ and $(k + 5)$

(B) $(k + 30)$ and $(k - 1)$ (D) $(k - 10)$ and $(k + 3)$

3. $(3x^2)^3 =$
(A) $3x^6$ (B) $9x^5$ (C) $27x^6$ (D) $27x^5$

4. Find the area of a rectangle whose length is $9x$ and width is $4x$.

5. The formula $d = rt$ gives you the distance traveled when you multiply the rate of the vehicle by the amount of time it traveled. A plane travels at a rate represented by $(y + 80)$. If it travels in $(4y - 3)$ hours, represent the distance the plane traveled.

6. The perimeter of an equilateral triangle is represented as $12x + 27$. Find the length of one side.

7. One factor of $15x^4 + 20x^3 - 35x^2$ is $5x^2$. Find the other factor.

8. The perimeter of a square is $4p - 12$. Express the area as a trinomial.

9. Find the area of a square whose side is $6x^2y^3$.

10. Factor $d^2 - 81$.

11. One factor of $x^2 - 14x + 48$ is $(x - 8)$. Find the other factor.

12. David factored $y^2 + 5y - 6$ and got $(y - 2)$ and $(y + 3)$, which is wrong. Explain to David why his answer is incorrect.

13. The area of a rectangle is $k^2 + 3k - 28$. Find its dimensions.

14. Find the factors of $x^2 - 11x + 24$.

15. Factor: $y^2 - 3y - 18$.

Chapter 7

Geometry

LINES

A *point* is a location in space. It has neither height nor length. Two points determine a *line*, which is a set of points that follow a path going indefinitely in both directions. A portion of a line that has two endpoints is a *line segment*. A *ray* is a section of a line that has only one endpoint. A line is identified by any two points on the line, but it can also be given a name using a single lowercase letter. Points are identified as uppercase letters. A line segment is identified by its two endpoints. A ray is identified by the endpoint, which is written first, and another point on the ray.

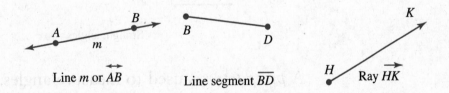

Line *m* or $\overleftrightarrow{AB}$ Line segment $\overline{BD}$ Ray $\overrightarrow{HK}$

Parallel lines are equidistant (the same distance) from each other and never intersect. Right angles are formed when two lines intersect and are perpendicular to each other. The symbol for a right angle is a small square at the vertex of the angle (the endpoint shared by the two segments that form the angle).

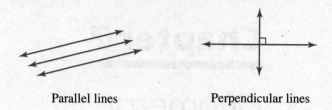

Parallel lines   Perpendicular lines

ANGLES

When segments, rays, or lines meet or intersect, angles are created. Angles whose measure is between 0° and 90° are called *acute angles*. A 90° angle is called a *right angle*. If the measure of an angle is between 90° and 180°, the angle is an *obtuse angle*. An angle whose measure is exactly 180° is called a *straight angle*.

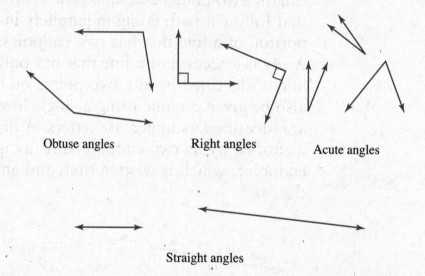

Obtuse angles   Right angles   Acute angles

Straight angles

A *protractor* is used to measure angles. There are two rows of numbers on a protractor. Place the crosshairs of the protractor on the vertex of the angle so that the horizontal line of the crosshairs lies on one of the rays of the angle. The other ray of the angle will fall on one of the numbers of the protractor. If the ray lands on the numbers 70° and 110°, the measure of the angle is 70° if the angle is acute or 110° if it is obtuse. The angle in the next diagram measures 110° because it is an obtuse angle.

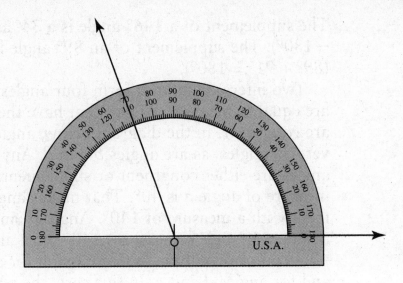

EXAMPLES

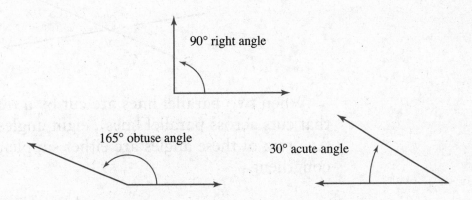

90° right angle

165° obtuse angle

30° acute angle

Use your protractor to verify these measurements. The rays that form angles continue indefinitely. Extend the rays if you need more length to measure an angle.

If the sum of two angles is 90°, the angles are *complementary angles*. Think of *c*omplementary and *c*orner. The angles form a right angle like the inside corner of a room. The complement of a 67° angle is a 23° angle (67° + 23° = 90°). The complement of a 12° angle is a 78° angle (12° + 78° = 90°).

If the sum of two angles is 180°, the angles are *supplementary angles*. Think of *s*upplementary and *s*traight. The angles form a straight angle that appears as a straight line.

The supplement of a 146° angle is a 34° angle (146° + 34° = 180°). The supplement of an 89° angle is a 91° angle (89° + 91° = 180°).

Two intersecting lines form four angles. *Vertical angles* are equal in measure. Angles that have the same measure are *congruent*. In the diagram below, angles *a* and *c* are vertical angles, as are angles *b* and *d*. Any two of these angles are either congruent or supplementary. Suppose the measure of angle *a* is 40°. That makes angle *b* its supplement with a measure of 140°. Angles *b* and *c* are supplementary. If angle *b* measures 140°, then angle *c* measures 40°. You can continue this discussion for angles *c* and *d* and for angles *d* and *a*. In this case, the pair of vertical angles *a* and *c* are 40°, and angles *b* and *d* are each 140°.

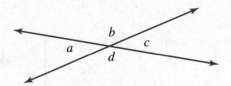

When two parallel lines are cut by a *transversal* (a line that cuts across parallel lines), eight angles are formed. Any two of these angles are either supplementary or congruent.

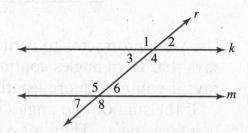

Imagine that you can place line *m* on line *k*. Then the pairs of angles 1 and 5, 2 and 6, 3 and 7, and 4 and 8 are called *corresponding angles* and are congruent.

Angles 3 and 6, and angles 4 and 5, are called *alternate interior angles*. They are on opposite sides of the transversal and are between the two parallel lines. These angles are congruent.

Angles 2 and 7, and angles 1 and 8, are called *alternate exterior angles,* and they also are congruent. These angles are on opposite sides of the transversal but lie outside the parallel lines.

If the measure of angle 3 is 30°, you can find the measure of angle 8. You know that angles 3 and 7 are corresponding angles and that angles 7 and 8 are supplementary. Thus, angle 8 measures 150° (180° − 30° = 150°).

PLANE FIGURES

Triangles

All triangles have three angles and three sides. An *equilateral triangle* has three congruent sides and three angles equal in measure. An *isosceles triangle* has two congruent sides and two congruent angles. A *scalene triangle* has three sides of different lengths.

By combining the terms *acute, obtuse,* and *right* with *equilateral, isosceles,* and *scalene,* we can describe any triangle.

EXAMPLES

Identify the triangles below based on their angles and sides.

A.

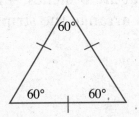

This is an equilateral triangle because all sides are congruent and all angles are equal in measure. All angles in an equilateral triangle are acute.

B.

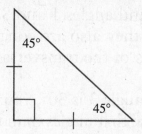

This is a right isosceles triangle because it has a right angle and two congruent sides. The two acute angles in a right isosceles triangle are congruent.

C.

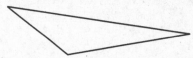

This is an obtuse scalene triangle because there are no congruent sides and the largest angle is obtuse.

D.

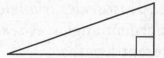

This is a right scalene triangle because there is a right angle and no congruent sides.

An interesting property of triangles is that the sum of the lengths of any two sides is always greater than the length of the third side. You can have a triangle whose sides are 4, 5, and 7 because (4 + 5) > 7, (4 + 7) > 5, and (5 + 7) > 4. You cannot have a triangle whose sides are 3, 4, and 10 because (3 + 4) < 10. Take thin strips of paper that are 3 inches, 4 inches, and 10 inches long. Then try to arrange the strips to create a triangle. It can't be done!

RIGHT TRIANGLES

Pythagoras was a Greek philosopher (c. 560–480 B.C.) who was the leader of a secret society that had many followers who valued knowledge. Some say that Pythagoras did not discover the theorem that is credited to him but, rather, that one of his followers discovered it. Whatever the case, the Pythagorean theorem is still used today.

If you square the lengths of the two legs of a right triangle, the square of the hypotenuse (the longest side) is equal to their sum. We know this as $a^2 + b^2 = c^2$, or as the *Pythagorean theorem.*

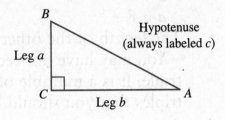

EXAMPLES

A. Find the length of the hypotenuse of a right triangle whose legs are 3 and 4.

$$a^2 + b^2 = c^2$$
$$3^2 + 4^2 = c^2$$
$$9 + 16 = c^2$$
$$25 = c^2$$
$$c = +5 \text{ or } -5$$

Since lengths are not negative, we disregard –5. We will consider only the positive value of the square root. The length of the hypotenuse is 5.

This grouping (3-4-5) is called a Pythagorean triple. Multiples of 3-4-5 are also Pythagorean triples, such as 9-12-15 and 15-20-25.

B. The hypotenuse of a right triangle is 10, and one of its legs is 6. What is the length of the other leg? Substitute the values into the formula.

$$a^2 + b^2 = c^2$$
$$a^2 + 6^2 = 10^2$$
$$a^2 + 36 = 100$$
$$a^2 = 64$$
$$a = 8$$

The length of the other leg is 8.

You may have noticed that 6-8-10 is a Pythagorean triple. It is a multiple of 3-4-5. Two other Pythagorean triples that you should know are 5-12-13 and 7-24-25.

C. Find the length of the hypotenuse when the legs of a right triangle have lengths of 7 and 10.

$$a^2 + b^2 = c^2$$
$$7^2 + 10^2 = c^2$$
$$49 + 100 = c^2$$
$$149 = c^2$$
$$c = \sqrt{149}$$

$\sqrt{149}$ is not a perfect square. You can leave your answer in this form or approximate it using your calculator.

$$\sqrt{144} < \sqrt{149} < \sqrt{169}$$
$$12 < \sqrt{149} < 13$$

$\sqrt{149}$ is between 12 and 13. To the nearest tenth, the length of the hypotenuse $\sqrt{149}$ is 12.2.

D. The hypotenuse of a right triangle is 35. One leg has a length of 20. Find the length of the other leg to the nearest tenth using your calculator.

$$a^2 + b^2 = c^2$$
$$a^2 + 20^2 = 35^2$$
$$a^2 + 400 = 1225$$
$$a^2 = 825$$
$$a = \sqrt{825}$$
$$a \approx 28.7$$

Using the Pythagorean theorem will help you determine if a triangle is a right triangle when given the lengths of all three sides.

E. The lengths of the sides of a triangle are 9, 17, and 20. Determine if this triangle is a right triangle.

$$a^2 + b^2 = c^2$$
$$9^2 + 17^2 = 20^2$$
$$81 + 289 = 400$$
$$370 \neq 400$$

This triangle is not a right triangle.

F. A triangle has sides of 18, 24, and 30. Is it a right triangle?

$$a^2 + b^2 = c^2$$
$$18^2 + 24^2 = 30^2$$
$$324 + 576 = 900$$
$$900 = 900$$

Yes, this triangle is a right triangle. Also notice that 18-24-30 is a multiple of the Pythagorean triple 3-4-5.

QUADRILATERALS

A *quadrilateral* is a four-sided plane figure. Quadrilaterals are classified by their attributes.

Parallelograms

If a quadrilateral has two pairs of opposite sides that are parallel, then the quadrilateral is a *parallelogram*. A rectangle has two pairs of parallel sides, four right angles, and congruent diagonals. A square is a special instance of a rectangle that has four congruent sides, four congruent right angles, and congruent diagonals. A rhombus is a parallelogram with four congruent sides.

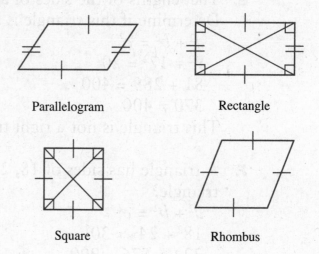

Parallelogram   Rectangle

Square   Rhombus

Trapezoids

If a quadrilateral has one pair of parallel sides, it is a trapezoid. Depending on how it is constructed, a trapezoid may or may not have a right angle. It will be an isosceles trapezoid where the nonparallel sides are equal in length and the base angles are equal.

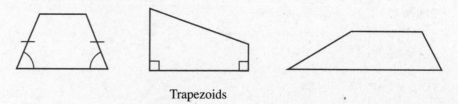

Trapezoids

OTHER POLYGONS

Poly- is Greek for "many" and *-gon* comes from the Greek word for knee, bend, or angle. Polygons get their names from the number of angles they have. Since they have the same number of angles as sides, we identify a polygon by the number of sides it has.

Number of Angles	Name of Figure
3	Triangle
4	Quadrilateral
5	Pentagon
6	Hexagon
7	Heptagon
8	Octagon
9	Nonagon
10	Decagon

Do your own research to find the names of polygons with more than ten sides.

If all the sides of a figure are congruent (the sides have the same length) and all the angles are congruent (the angles are equal in measure), then the figure is called a *regular* polygon. A regular triangle is called an equilateral triangle. A regular quadrilateral is called a square. From then on, figures have the word *regular* in front of their names, such as regular octagon, when all of their sides and their angles are congruent.

INTERIOR ANGLES OF POLYGONS

The interior angles of a triangle measure 180°. Trace the triangle below on a sheet of paper. Then carefully tear the triangle in three pieces so the angles are separated.

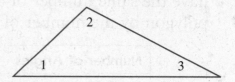

Reassemble the angles so the vertices meet at the same point and the pieces do not overlap. Together the angles form a straight angle. Their sum is 180°.

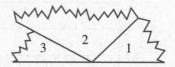

A rectangle, like a square, has four angles that each measure 90°. What about quadrilaterals that have angles not equal to 90°? In the diagram below, lines k and m are parallel, as are lines p and r. They form a parallelogram with two acute angles, a and f, and two obtuse angles, e and c. Earlier in this chapter, the relationships between angles formed when parallel lines are cut by a transversal were discussed. Because angles a and b are supplementary and angles b and c are congruent, angles a and c are also supplementary. The same can be said for angles e and f. Thus, the sum of the interior angles of a quadrilateral is 2(180°) or 360°.

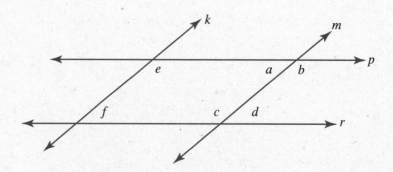

The pentagon below is composed of a square and an equilateral triangle.

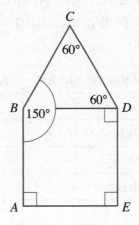

Angles A and E each measure 90°. Angles B and D each measure 150° (90° + 60° = 150°). The sum of the interior angles of pentagon $ABCDE$ is 2(90°) + 2(150°) + 60° = 540°.

A regular hexagon is formed from six equilateral triangles. The interior angles of each triangle measure 180°, and each angle measures 60°. One interior angle of a hexagon measures 60° + 60° = 120°. The sum of the interior angles of a hexagon is 720°.

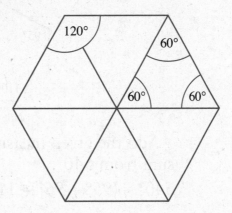

The table below shows our findings along with the measures of other figures. After studying the data, a pattern emerges. To find the measure of the interior angles of any polygon, multiply 180 by 2 less than the number of sides.

Name of Figure	Number of Sides	Sum of Interior Angles
Triangle	3	$180(3 - 2) = 180°$
Quadrilateral	4	$180(4 - 2) = 360°$
Pentagon	5	$180(5 - 2) = 540°$
Hexagon	6	$720°$
Heptagon	7	$900°$
Octagon	8	$1080°$
n-gon	n	$180°(n - 2)$

EXAMPLES

A. Find the missing angle of the pentagon.

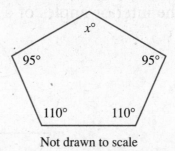

Not drawn to scale

Add the given measurements and then subtract the sum from 540°.

$540° - (95° + 95° + 110° + 110°) = 540° - 410° = 130°$

B. Find the missing angle.

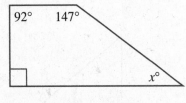

Not drawn to scale

The sum of the interior angles of a quadrilateral is 360°. There is one right angle that measures 90°.
$360° - (90° + 92° + 147°) =$
$360° - 329° = 31°$
So the measure of angle x is 31°.

SIMILARITY AND CONGRUENCE

Two figures are congruent if they have the same size and shape. Congruent line segments have the same length. Congruent angles have the same measure in degrees. All right angles are congruent because every right angle measures 90°. The symbol ≅ means *congruent*.

Two triangles are similar if their corresponding angles are congruent and the lengths of corresponding sides have the same ratio. Triangle ABC has sides with lengths of 3, 4, and 5 units. Triangle $A'B'C'$ has sides with lengths of 6, 8, and 10 units. Angles A and A' are 90°, $\angle B \cong \angle B'$, and $\angle C \cong \angle C'$. $\overline{AB}$ and $\overline{A'B'}$ are corresponding sides. Their ratio is 1 to 2. The ratio of the other pairs of corresponding sides is also 1 to 2. The symbol ~ means *similar*.

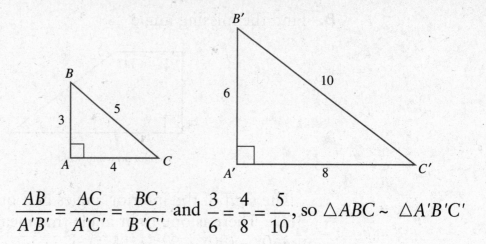

$$\frac{AB}{A'B'} = \frac{AC}{A'C'} = \frac{BC}{B'C'} \text{ and } \frac{3}{6} = \frac{4}{8} = \frac{5}{10}, \text{ so } \triangle ABC \sim \triangle A'B'C'$$

EXAMPLES

A. $\triangle FGH \sim \triangle KLM$. Find the lengths of the missing sides of $\triangle KLM$.

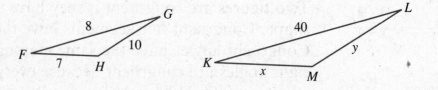

$\dfrac{FG}{KL} = \dfrac{8}{40} = \dfrac{1}{5}$ The ratio is $\dfrac{1}{5}$.

$\dfrac{FH}{KM} = \dfrac{7}{x} = \dfrac{1}{5}$ So the length of $\overline{KM}$ is 35 units.

$\dfrac{GH}{LM} = \dfrac{10}{y} = \dfrac{1}{5}$ Thus, the length of $\overline{LM}$ is 50 units.

B. In △*PQR*, $\overline{PQ} \cong \overline{QR}$ and ∠*P* ≅ ∠*R*. If ∠*P* measures 66°, find the measure of ∠*Q*.

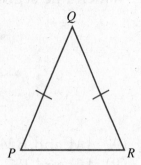

The sum of the interior angles of a triangle is 180°. We have the measure of the two base angles that are congruent. Find the measure of the third angle, *Q*.

$$2(66°) + Q = 180°$$
$$132° + Q = 180°$$
$$Q = 48°$$
$$\text{So } \angle Q = 48°$$

TEST YOUR SKILLS   (See page 170 for answers.)

1. Which line segments form angle *k*?

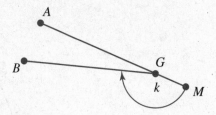

    (A) *AM* and *BM*
    (B) *AG* and *BG*
    (C) *MG* and *BG*
    (D) *MG* and *AG*

2. Given: $QT = 12$, $WT = 20$, $QW = 16$, and $\angle Q$ is a right angle. Which of the following statements is true?

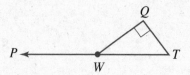

    (A) Triangle *WQT* is an isosceles triangle.
    (B) Angles *QWT* and *QTW* are supplementary.
    (C) Angles *PWT* and *QTW* are supplementary.
    (D) Angles *QWT* and *QTW* are complementary.

3. Given that the measure of $\angle 4$ is 127°, what is the sum of $\angle 6$ and $\angle 7$?

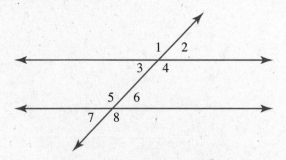

    (A) 53°   (C) 180°
    (B) 106°   (D) cannot be determined

4. You are given sticks with lengths of 2, 7, and 9 centimeters to make a triangle. Using any segment as many times as you wish, answer the following questions. Explain your answer in each part.

   A. How many equilateral triangles can be constructed with these segments?

   B. How many isosceles triangles can be constructed?

   C. How many scalene triangles can be constructed?

5. $\triangle ABC \sim \triangle CDE$, and $\angle ACD$ and $\angle BCE$ are straight angles. Give two reasons why $\angle x \cong \angle y$.

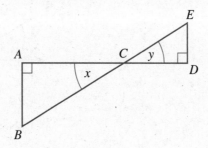

6. You are told that $\triangle ABC \sim \triangle DEF$. Find the lengths of $\overline{ED}$ and $\overline{FD}$. Show your work.

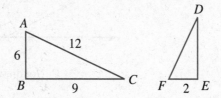

7. Choose the one correct statement below and explain why it is true.

   (A) All vertical angles are congruent.

   (B) All trapezoids are parallelograms.

   (C) The complement to an angle of 86° is 94°.

8. Use the accompanying figure to answer the questions.
   A. Name one pair of congruent angles.
   B. Name one pair of supplementary angles.

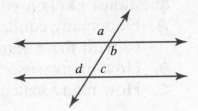

9. $\angle r = x + 3$ and $\angle s = 2x + 6$
   What is the measurement of $\angle s$?

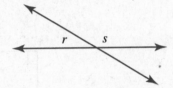

Not drawn to scale

10. In the diagram below, find the measurement of $\angle z$ if $\angle x$ is half the measurement of $\angle z$.

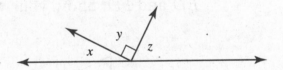

Not drawn to scale

11. $\angle p + \angle r + \angle s = 180°$. Find the measurement of $\angle t$.

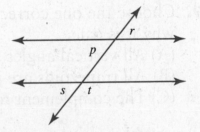

Not drawn to scale

12. In the diagram, three line segments intersect and form △XYZ. If ∠b = 35° and ∠c = 80°, find the measure of ∠Y.

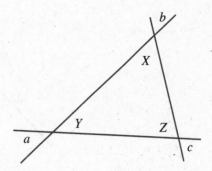

13. Find the measure of ∠3 if ∠4 = 3x + 2 and ∠7 = x − 10.

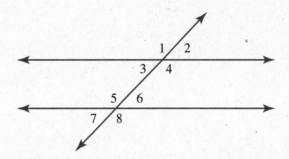

14. In the diagram, △ABC is a right isosceles triangle. Find the measure of ∠x.

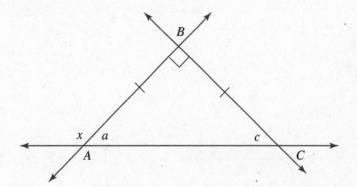

15. In the diagram below, two parallel lines are cut by a transversal. Find the measures of angles a and b if $\angle a = 3x - 5$ and $\angle b = 10x + 3$.

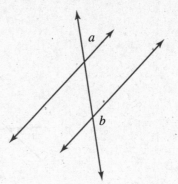

Chapter 8

Graphing

THE COORDINATE SYSTEM

The Cartesian coordinate system is named after René Descartes, a famous French mathematician and philosopher. Two perpendicular lines intersect at the *origin* and divide the plane into four *quadrants*. The horizontal line is called the *x-axis*, the number line with which you are familiar, and the vertical number line is called the *y-axis*. The direction of the numbers on the y-axis is similar to that of a thermometer, with the numbers decreasing as the temperature falls. The coordinates of the origin are (0, 0). The diagram shows the signs of the coordinates in each of the quadrants. The first number in an ordered pair is the *x-coordinate*, and the second is the *y-coordinate*. Order matters in the pair. The point (5, 2) is *not* the same point as (2, 5).

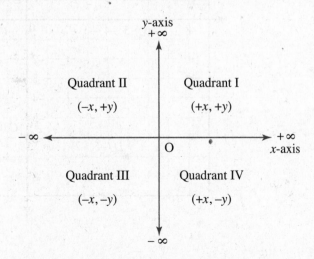

PLOTTING POINTS

To plot points, find the location of the first number of the ordered pair on the (horizontal) x-axis. From there, move the number of units on the (vertical) y-axis according to the second number, going up if the number is positive or down if the number is negative.

EXAMPLE

Plot and label the following points on the grid below.

A (4, 4) B (–7, 6) C (0, –6)
D (–5, –7) E (7, 0) F (2, –4)

Locate the first number on the x-axis. From this location, find the second number along the y-axis.

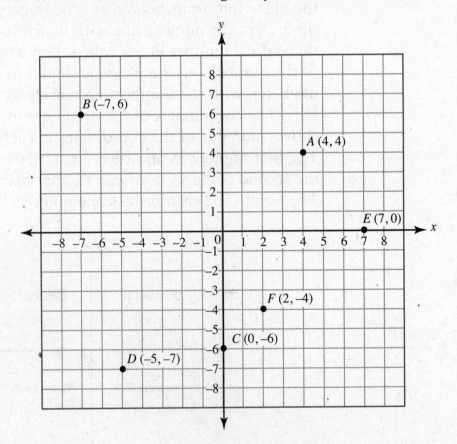

SLOPE

The *slope* of a line indicates how steep it is. Slope is based on the ratio of the vertical change over the horizontal change in the line as you move one point to another. Follow the line from left to right. If the line rises (⤢), the slope is positive. If it falls (⤡), the slope is negative. Horizontal lines (↔) neither rise nor fall and have a slope of zero. Vertical lines (↕) have a slope that is undefined. A positive slope has a ratio $\frac{\text{rise}}{\text{run}}$, and a negative slope has a ratio $\frac{\text{fall}}{\text{run}}$. The direction or the run always goes from the left toward the right.

EXAMPLES

A. Find the slope of the line below. The scale of the graph is one unit.

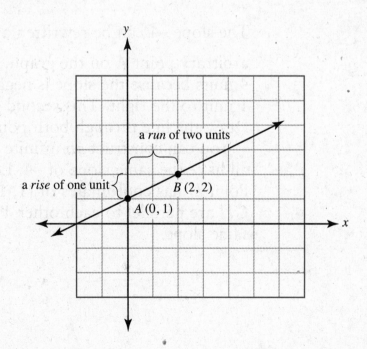

Locate two points on the line. To get from *A* to *B*, go up 1 unit and go to the right 2 units. The rise is 1, and the run is 2. The slope of this line is $\frac{1}{2}$. Because the line rises as you follow it from the left to the right, the slope is positive. No matter which two points you choose on this line, the rate of change (rise of 1 unit and run of 2 units) will be the same as you move from left to right.

B. Draw a line that has a slope of −4.

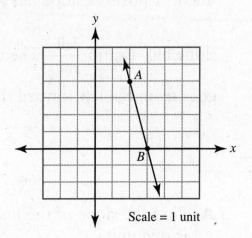

Scale = 1 unit

The slope −4 can be rewritten as $\frac{-4}{1}$. Locate an arbitrary point *A* on the graph. From *A*, count down 4 units because the slope is negative and then move 1 unit to the right. This second point on the line is *B*. Draw the line through both points.

You can construct an infinite number of lines that all have the same slope of −4. Locate point *C* at (0, 5). Point *D* has coordinates of (1, 1). Lines *CD*, *EF*, and *GH* are parallel to each other. Parallel lines have the same slope.

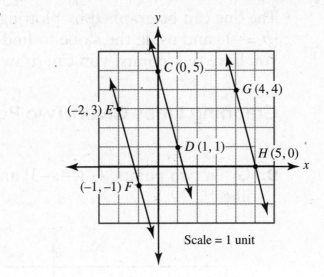

Scale = 1 unit

GRAPHING LINES

Slope-Intercept Form of a Linear Equation

The slope-intercept form of the equation of a line is $y = mx + b$, where m is the slope of the line and b is the y-intercept. The y-intercept is where the line hits the y-axis. The coordinates of the points on the line are the x- and y-values. If you are given an equation that is not in this form, rewrite the equation using your algebra skills. Once in slope-intercept form, you can easily graph the line.

EXAMPLE

A. Rewrite this linear equation in slope-intercept form:
$$4y + 6 = 2 + 3x$$

$$4y + 6 = 2 + 3x$$
$$4y = -4 + 3x$$
$$4y = 3x - 4$$
$$y = \frac{3}{4}x - 1$$

The line can be graphed by plotting the y-intercept (*b* = –1) and using the slope to find a second point. Once you have two points, you can draw the line.

Graphing Lines Given Two Points

EXAMPLES

B. Given two points, G (–4, –3) and S (2, 4), graph the line.

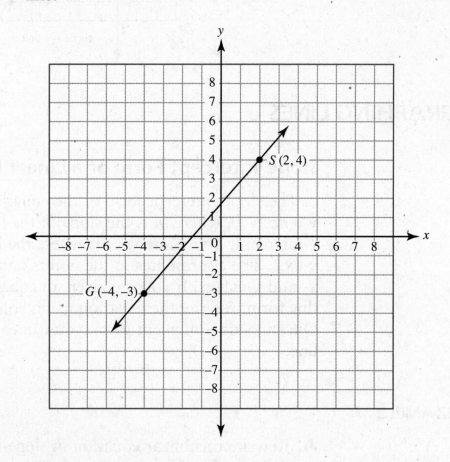

C. Given the y-intercept of 5 and point K (7, –3), graph the line.

The point where the line passes through the y-axis is called the *y-intercept*. The x-coordinate of this point will always be 0. The coordinates of the y-intercept of 5 are (0, –5). Call this point M. Plot point K then draw the line.

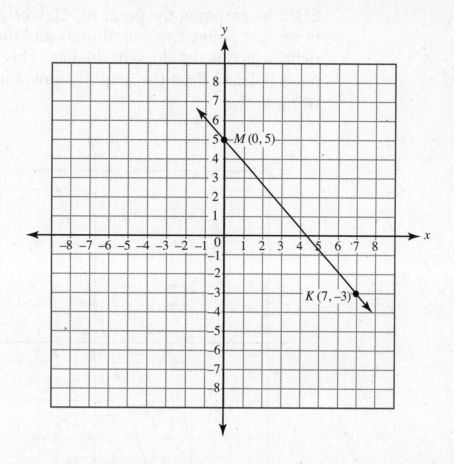

Graphing Lines Given One Point and the Slope

Equations of lines can be written using the form $y = mx + b$, which is also called the *slope-intercept* form. The variable m is the slope of the line, and b is the y-intercept or the point where the line intersects the y-axis. The x and y are the coordinates for any point on the line. When written in this form, you can get the data needed to graph a line.

EXAMPLES

D. Graph the line $y = 3x + 2$.

This equation is already written in the slope-intercept form. We know that this line passes through the y-axis at the point $(0, 2)$ and has a positive slope of 3.

Begin by graphing the point (0, 2), labeled *A* on the graph. From that point, move up 3 units and then 1 unit to the right, which is the slope of the line. This is your second point (labeled *B* on the graph). Draw a line through the two points.

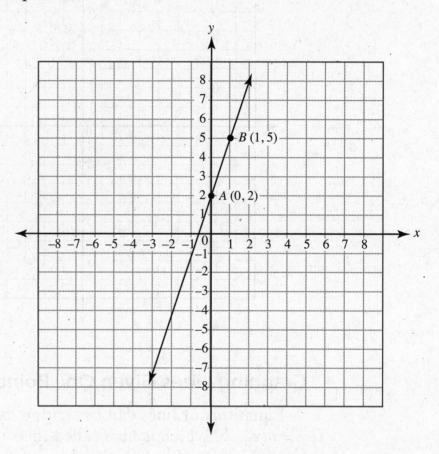

E. Graph the line $4x - 3y = 6$.
Rewrite the equation in slope-intercept form.

$$4x - 3y = 6$$
$$4x + (-4x) - 3y = (-4x) + 6$$
$$-3y = -4x + 6$$

$$\frac{-3y}{-3} = \frac{-4}{-3}x + \frac{6}{-3}$$

$$y = \frac{4}{3}x - 2$$

The slope of this line is $\frac{4}{3}$, and the y-intercept is –2. First locate –2 on the y-axis. This is the y-intercept. Label it point B (0, –2). From this point, count up 4 units, then right 3 units. Label this new point A (3, 2). Draw the line through the two points.

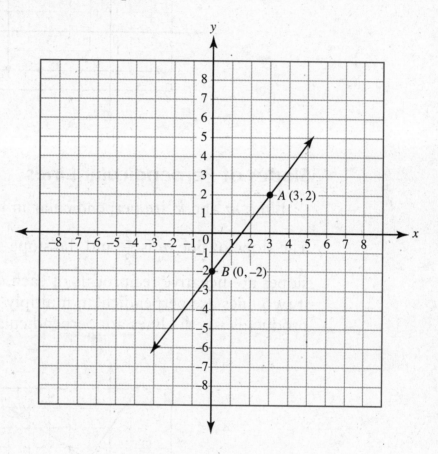

Equations of Vertical and Horizontal Lines

Vertical and horizontal lines have special equations. In the figure on the next page, the vertical line has the equation $x = 5$ because the x-coordinate of every point that lies on this line is 5. The equation of the horizontal line is $y = -2$. Every point on this line has a y-coordinate of –2. Generally speaking, vertical lines have the equation $x = a$, where a is the x-intercept. Horizontal lines have the equation $y = b$, where b is the y-intercept.

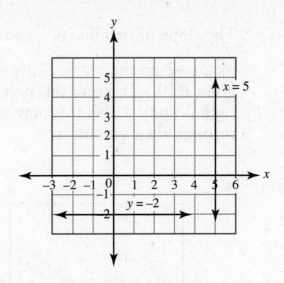

Slopes of Perpendicular Lines

Lines *m* and *k* are perpendicular in the example below.

The slope of line *m* is $\frac{2}{3}$ and the slope of line *k* is $\frac{-3}{2}$. The slopes are negative reciprocals of each other. To determine if two lines are perpendicular, multiply their slopes. If the product is –1, the lines are perpendicular.

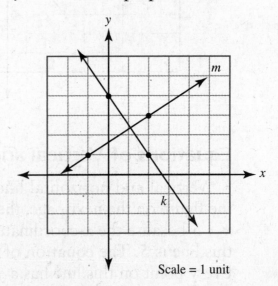

Scale = 1 unit

Determining if a Point Is on a Line

You can determine if a point lies on a given line by substituting the x- and y-coordinates in the equation of the line. If your statement is true, the point lies on the line. If it is false, the point is not on the line.

EXAMPLES

F. Does the point (4, 3) lie on the line $y = 2x - 5$?
Substitute 4 for x and 3 for y in the equation.

$$y = 2x - 5$$
$$3 \overset{?}{=} 2(4) - 5$$
$$3 \overset{?}{=} 8 - 5$$
$$3 = 3$$

Yes, (4, 3) does lie on the line.

G. Does the point (3, 11) lie on the line whose equation is $y = 5x + 2$?
Substitute 3 for x and 11 for y.

$$11 \overset{?}{=} 5(3) + 2$$
$$11 \overset{?}{=} 15 + 2$$
$$11 \neq 17$$

No, (3, 11) is not on the line.

H. List three points that lie on the line $8x + 2y = 10$.
Make a table of values by choosing at least three values for x. Two points may determine a straight line, but having more than two points ensures a straight line. You should include a positive value, a negative value, and 0. Then find the y-coordinate by solving the equation.

x-coordinate	y-coordinate	Point
−1	$8(-1)+2y=10$ $-8+2y=10$ $2y=18$ $y=9$	(−1, 9)
0	$0+2y=10$	(0, 5)
1	$8+2y=10$ $2y=2$	(1, 1)

The points (−1, 9), (0, 5), and (1, 1) are on the given line.

I. Graph the line $y = x$.

Each point that lies on this line has the same value for both coordinates. Some points on this line are (−2, −2), (0, 0), and (4, 4). Notice that the slope, m, is understood to be 1. There is no y-intercept stated, so $b = 0$ and the line passes through the origin (0, 0).

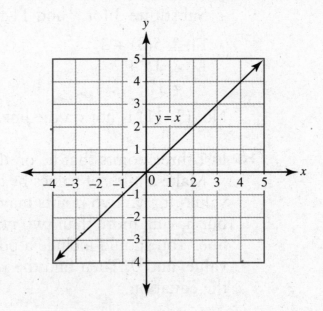

WRITING EQUATIONS OF LINES

To write the equation of a line, begin with the slope-intercept formula $y = mx + b$. Substitute the slope for m, and substitute the x- and y-coordinates to solve for b. Then substitute the m and b in the slope-intercept formula.

EXAMPLES

A. Write the equation of the line that has a slope of $\frac{1}{2}$ and that passes through the point (6, 1).

Substitute $\frac{1}{2}$ for m, 6 for x, and 1 for y. Then solve for b.

$$y = mx + b$$
$$1 = \frac{1}{2}(6) + b$$
$$1 = 3 + b$$
$$-2 = b$$

Now that you have the y-intercept, substitute m and b in the formula to get the equation of the line with a slope of $\frac{1}{2}$ and that passes through the point (6, 1).

$$y = \frac{1}{2}x - 2$$

B. Write the equation of a line that passes through the origin and has a slope of –3.

The origin is the same as the point (0, 0), where the x- and y-axes intersect. Substitute –3 for the slope into the slope-intercept formula. Since b is 0, you do not need to write it.

So the equation of the line that has a slope of –3 and passes through the origin is $y = -3x$.

Earlier in this chapter, slope was described as $\dfrac{\text{rise}}{\text{run}}$ for positive slopes or $\dfrac{\text{fall}}{\text{run}}$ for negative slopes, always following the line from left to right. This *rise* or *fall* is a change between the y-coordinates of two points on the line. The *run* is the change between the x-coordinates of those same two points. In previous examples, you counted the changes and represented the slope as a rational number. The formula for finding the slope of a line when you are given two points (x_1, y_1) and (x_2, y_2) is

$$m = \frac{y_2 - y_1}{x_2 - x_1}$$

Be sure to keep the coordinates in their correct place. Do not mix them when substituting in this formula. Once you have the slope, find *b* by substituting *m* and the coordinates of one of the points into the slope-intercept equation.

EXAMPLE

C. Write the equation of a line that passes through (–1, –5) and (2, 1).

Use the equation to find the slope. Let (–1, –5) be the first point and (2, 1) the second point. It does not matter which point is used as the first point and which is used as the second point. You must not mix up the values. The subscripts of 1 and 2 will help you substitute the correct value.

$$m = \frac{y_2 - y_1}{x_2 - x_1} = \frac{1 - (-5)}{2 - (-1)} = \frac{6}{3} = 2$$

Now substitute the coordinates of one point and m into the slope-intercept form to solve for b. Which point you use does not matter.

$$y = mx + b$$
$$1 = 2(2) + b$$
$$1 = 4 + b$$
$$-3 = b$$

Use the slope-intercept formula again. The equation of a line that passes through $(-1, -5)$ and $(2, 1)$ is $y = 2x - 3$.

SOLVING SYSTEMS OF EQUATIONS

When two straight lines intersect, three outcomes are possible.

1. The lines do not intersect. They are parallel.

2. The lines lie on top of each other.

3. The lines intersect at one point.

We tend to examine lines on a grid near the origin. Just because lines do not intersect in the small section we see on a grid does not mean that they do not intersect elsewhere. Remember that lines continue into infinity. The only lines that never intersect are parallel lines. Parallel lines have the same slope. Do not rely on what you see on the graph to determine if two lines are parallel. Rely on their slopes. For example, $y = 2x + 85$ and $y = x + 79$ do not intersect near the origin. Since they have different slopes, though, they do eventually intersect at $(-6, 73)$. In contrast, $y = 2x + 85$ and $y = 2x + 79$ never intersect because they have the same slope, 2.

Two lines can occupy the same location on a grid. For example, $y = 2x + 1$ and $3y = 6x + 3$ describe the same set of points. When graphed, there appears to be only one line because the lines lie on top of each other.

Two lines will intersect each other at one point. These lines have different slopes and do not lie on top of each other.

EXAMPLES

A. Find a pair of numbers whose sum is 16 and difference is 2.

Let x represent the first number and y represent the second. Write an equation that describes each situation.

The sum of the two numbers is 16: $x + y = 16$
The difference of the numbers is 2: $x - y = 2$

Rewrite each equation using the slope-intercept form of the equation of a line.

$$y = -x + 16 \qquad \text{and} \qquad y = x - 2$$

Graph each line on the same grid. The point where the lines intersect is the solution to this system of equations. Be sure to label the lines with their respective equation.

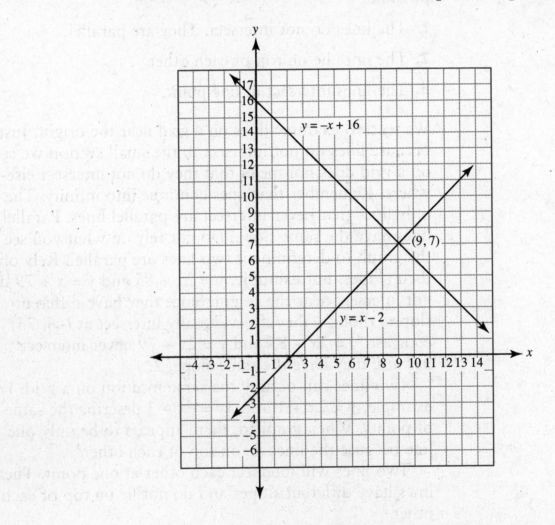

The point (9, 7) is on both lines. It is the solution to this system of equations because the sum of the numbers is 16 and their difference is 2.

B. Graph the lines to find the solution to this system of equations: $x + 2y = 7$ and $y = 2x + 1$.

Rewrite any equation that needs to be in the slope-intercept form. Then graph to find the point of intersection.

$$x + 2y = 7$$
$$2y = -x + 7$$
$$y = \frac{-1}{2}x + \frac{7}{2}$$

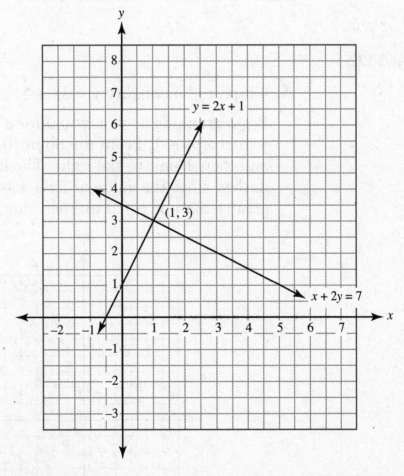

The solution to this system of equations is the point (1, 3).

GRAPHING LINEAR INEQUALITIES

Recall that an inequality is a statement that uses the signs <, ≤, >, ≥, or ≠. Graphing inequalities is similar to graphing linear equations. The difference is that the solution set is shaded on the grid and the line can be solid or dashed. The table below will help you with graphing linear inequalities.

If the Inequality Begins With	The Line Is	The Shaded Area Is
$y <$	Dashed	Below the line
$y >$	Dashed	Above the line
$y \leq$	Solid	Below the line
$y \geq$	Solid	Above the line

EXAMPLES

A. Graph the inequality $y < 3x + 1$.

Begin graphing as you would for a linear equation. The y-intercept is 1. From the point (0, 1), count 3 units up and then 1 unit to the right. The line you draw will be dashed since the value of $3x + 1$ is greater than, not greater than or equal to, the value of y. Watch the sign.

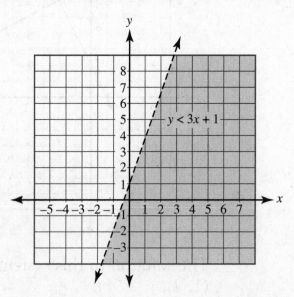

Since the y-values are less than $3x + 1$, shade the area below the line.

B. Graph the inequality $y \geq x + 5$.

The graphed line will be solid, and the shaded area will be above the line.

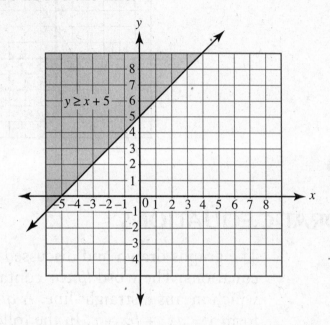

C. Graph the inequality $3x + 8 \geq 4 + 2y$.

Rewrite the inequality in the slope-intercept form and then graph it.

$$3x + 8 \geq 4 + 2y$$
$$3x + 4 \geq 2y$$
$$\frac{3}{2}x + 2 \geq y \quad \text{or} \quad y \leq \frac{3}{2}x + 2$$

The line will be solid, and the shaded area will be below the line.

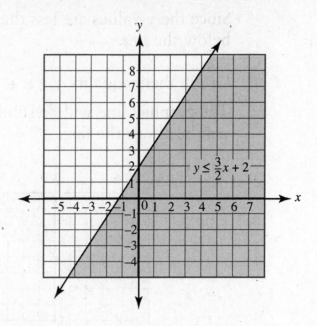

$y \le \frac{3}{2}x + 2$

QUADRATIC EQUATIONS

The graphs drawn and discussed so far have been linear equations. The word *linear* contains the root word *line*, which means a straight line. A *quadratic equation* is of the form $y = ax^2 + bx + c$. In the following examples, the quadratic equations will have $a = 1$. The graph of a quadratic equation is a *parabola*. The graph curves up and looks like this ∪ when the sign on the x^2 term is positive. When the sign on the x^2 term is negative, the graph will curve down and look like this ∩.

To graph a quadratic equation, create a table of values based on the equation just as you would for a linear equation. On the next page is the table for the quadratic equation $y = x^2 - 2x + 1$.

x	$y = x^2 - 2x + 1$	y	Point
–2	$y = (-2)^2 - 2(-2) + 1$ $y = \ 4 \ + \ \ 4 + 1$	9	(–2, 9)
–1	$y = (-1)^2 - 2(-1) + 1$ $y = \ 1 \ + \ \ 2 + 1$	4	(–1, 4)
0	$y = (0)^2 - 2\,(0) + 1$ $y = \ 0 \ - \ \ 0 + 1$	1	(0, 1)
1	$y = (1)^2 - 2(1) + 1$ $y = \ 1 \ - \ \ 2 + 1$	0	(1, 0)
2	$y = (2)^2 - 2(2) + 1$ $y = \ 4 \ - \ \ 4 + 1$	1	(2, 1)
3	$y = (3)^2 - 2(3) + 1$ $y = \ 9 \ - \ \ 6 + 1$	4	(3, 4)
4	$y = (4)^2 - 2(4) + 1$ $y = 16 \ - \ \ 8 + 1$	9	(4, 9)

The graph of $y = x^2 - 2x + 1$ will curve up. Plot your points, and connect them with a slightly curved line.

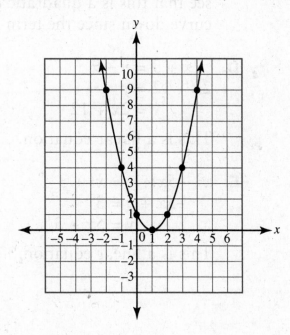

EXAMPLES

Classify the equations below as either linear or quadratic. If quadratic, determine if the shape curves up or down.

A. $y = x^2 - 4x + 3$

The equation is in the standard quadratic form. This quadratic will curve up since the term x^2 is positive.

B. $y = 7x - 2$

The exponent on the highest term is 1. This is a linear equation.

C.
$$3x + 7 = 5 - x^2 + y$$
$$3x + 2 = -x^2 + y$$
$$3x + 2 - x^2 = y$$
$$y = -x^2 + 3x + 2$$

Now that the equation is in standard form, you can see that this is a quadratic equation whose shape will curve down since the term x^2 is negative.

D.
$$2x + 3 = y - 9$$
$$2x + 12 = y$$
$$y = 2x + 12$$

This is a linear equation.

E.
$$x^2 - 5x + y = x^2 + 3$$
$$-5x + y = 3$$
$$y = 5x + 3$$

This is a linear equation.

RELATIONS AND FUNCTIONS

In Chapter 3, you read that a function is like a machine. You put in a number, x, and a number comes out, y. The numbers inserted into the function machine form a set called the *domain*. The numbers that come out of the machine form a set called the *range*. The numbers in the domain are the x-coordinates or *independent variables*. The numbers in the range are the y-coordinates or *dependent variables* because the value of y depends on the value of x. The set of ordered pairs (x, y) is called a *relation*. If a relation has a rule where every element in the domain is paired with exactly one element in the range, then the relation is a *function*.

EXAMPLE

A. Use a table to represent the perimeters of equilateral triangles that have a side length in whole numbers greater than 3 and less than 8. The relation between the side length and perimeter is $y = 3x$.

Side Length (x)	Perimeter $(y$ or $f(x))$	Point (x, y)
4	12	(4, 12)
5	15	(5, 15)
6	18	(6, 18)
7	21	(7, 21)

In the above example, the domain is the set {4, 5, 6, 7} and the range is the set {12, 15, 18, 21}. The column showing the points has each element in the domain paired with exactly one element in the range. This relation is a function.

Both the domain and range can be written as inequalities. In the above example, the domain is $4 \leq x \leq 7$ and the range is $12 \leq y \leq 21$, where x and y are elements of $\mathbb{W}$.

The relation {(0, 4), (1, 5), (2, 6)} is a function since a pattern can be found and described as $y = x + 4$. The relation {(−1, 4), (−2, 8), (−3, 12), (−2, −8)} is not a function. The number −2 is paired with both 8 and −8.

You can determine if a graph is a function by checking to see that every element in the domain (on the x-axis) is paired with exactly one element in the range (y-axis). The simplest way to check is to use the *vertical line test*. Hold your pencil parallel to the y-axis and move it from left to right. If the graph, whether straight or curved, hits the pencil more than once, it fails the test and the graph is not a function.

EXAMPLES

B. Determine which graph is a function and explain your reasoning.

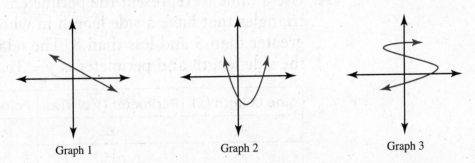

Graph 1   Graph 2   Graph 3

In Graph 1, you can easily see that every x is paired with one unique y. The vertical line test shows the same conclusion, that this is a function.

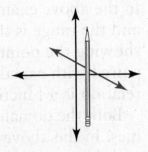

The vertical line test shows that Graph 2 is also a function.

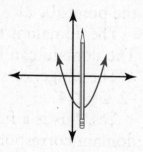

The vertical line test shows that Graph 3 is not a function. The left edge of the pencils hits the graph of the curved line three times, as shown by the circles. This means for one value of x, there are three values of y.

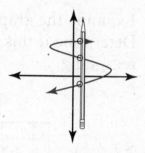

C. Examine the graph. State the domain. State the range. If this is a function, write the rule.

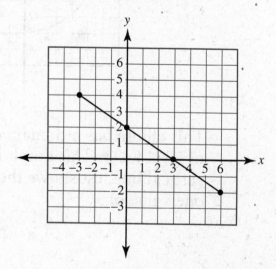

The endpoints of this line segment have coordinates of (−3, 4) and (6, −2). The line segment passes through the point (0, 2).

The domain is the set of x-values of this line segment. The domain can be written as −3 ≤ x ≤ 6.

The range, or set of y-values, can be written as −2 ≤ y ≤ 4.

Yes, this is a function since every element in the domain corresponds to only one element in the range. You can check this using the vertical line test. The rule of this function can be written as

$$y = \frac{-2}{3}x + 2.$$

D. Examine the graph. State the domain. State the range. Determine if this graph is a function and explain your reasoning.

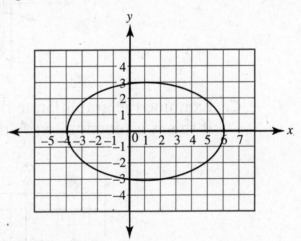

This ellipse has a domain of −4 ≤ x ≤ 6 and a range of −3 ≤ y ≤ 3. This graph is not a function because the vertical line test shows there are two values of y for one value of x.

DESCRIPTIVE GRAPHS

A graph can be used to describe a situation that shows a relationship between two things. The graph below shows a relationship between the distance from home and the time that has passed. Jonathan rode his bike to his friend's house, played video games, and then rode his bike home.

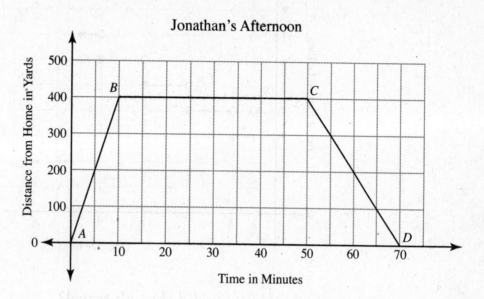

Jonathan starts out from home (*A*). He takes 10 minutes to ride 400 yards, or about a quarter mile, to his friend's house (*B*). Jonathan spends 40 minutes playing video games (*B* to *C*). He rides back home at a slower rate that takes him 20 minutes (*D*).

EXAMPLES

A. The graph below describes Shari's cab ride by showing the cost of the trip relative to distance. Answer questions A–C based on the graph.

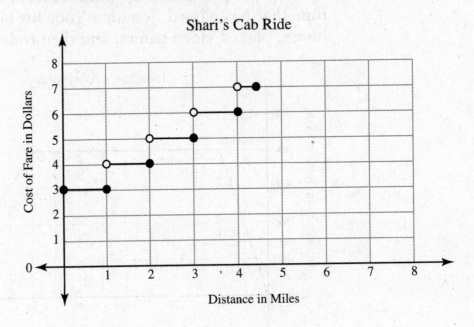

Shari's Cab Ride

A. How far did the cab travel?
The trip was 4.5 miles.

B. What was the cost of the trip?
The fare was $7.

C. What were the rates for the cab fare?
Shari paid $3 for the first mile and then $1 for each additional mile.

B. For your aunt's birthday, you bought her a bouquet of fresh flowers. She filled the vase with water and then added the flowers. The following graph shows the height of the water in the vase with respect to time. Study the graph and then choose which shaped vase was used. Explain your reasoning.

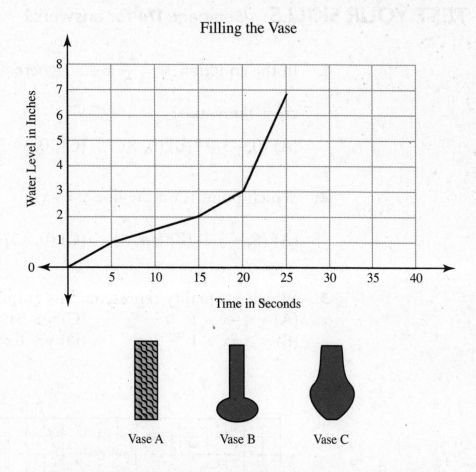

Filling the Vase

The graph shows the water level filling at the rate of 1 inch per 5 seconds, then slowing down to 1 inch per 10 seconds, then filling quicker at the rate of 1 inch per 5 seconds, and finally filling 4 inches per 5 seconds. Vase *B* was used. The bulbous bottom would fill slower toward the widest part of the bottom and then fill quickly at the top. Vase *A* would fill at a steady rate from beginning to end and its graph would be a straight line. The shape of Vase *C* only slightly changes, so the graph would reflect a gentle increase and decrease.

TEST YOUR SKILLS   (See page 174 for answers.)

1. In the equation $y = \dfrac{-3}{2}x + 8$, where does the line cross the y-axis?

   (A) $(0, -3)$ (B) $(0, 8)$ (C) $(0, y)$ (D) $(0, \dfrac{-3}{2})$

2. Which point is on the line $y = -x$?

   (A) $(6, \dfrac{1}{6})$ (B) $(-4, 4)$ (C) $(0, -1)$ (D) $(8, 8)$

3. Which inequality represents this graph?
   (A) $y \leq -x - 1$ (C) $y > 3x - 1$
   (B) $y \geq 6x + 1$ (D) $y > 6x - 1$

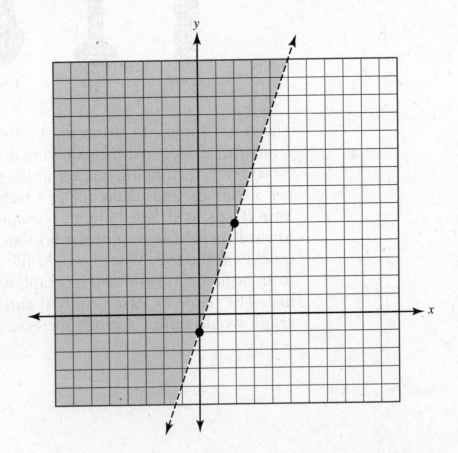

4. Which point does not lie on the line $y = \frac{1}{2}x - 1$?

   (A) $(-6, -4)$ (C) $(1, -0.5)$
   (B) $(0, -1)$ (D) $(6, 4)$

5. The equation of a line is given as $7x - y = 4$. Which is the equivalent equation written in slope-intercept form?
   (A) $-y = 7x + 4$ (C) $y = -7x - 4$
   (B) $-y = 4 + 7x$ (D) $y = 7x - 4$

6. Write the equation of a line that has a slope of 4 and that intersects the y-axis at $(0, 7)$.

7. The area of a rectangle is 6 units². Three of the vertices are $(1, -2)$, $(4, -2)$, and $(1, -4)$. What are the coordinates of the fourth vertex?

8. What is the slope of a line that is perpendicular to the line $y = \frac{-1}{2}x + 8$?

9. **A.** Make a table showing five points that lie on the line $y = 3x - 2$.

   B. Graph the line.

10. One leg of an isosceles right triangle is parallel to the x-axis, and the other leg is parallel to the y-axis. The area of this triangle is 18 units². Two vertices are located at $(5, 3)$ and $(-1, -3)$. What are the coordinates of the third vertex?

11. Find the solution graphically to this system of equations.

    $$2y - 6 = x \qquad \text{and} \qquad 6x + y = 3$$

12. Your dad purchased a small, round, above ground backyard pool. It is 2 feet deep with a diameter of 8 feet. Which graph depicts the water level while a pool is filling at a constant rate over a period of time?

(A)

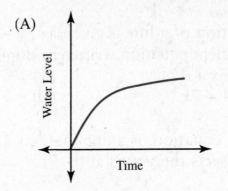

(B)

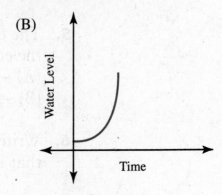

(C)

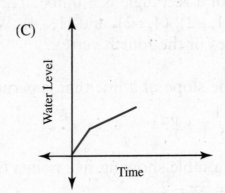

(D)

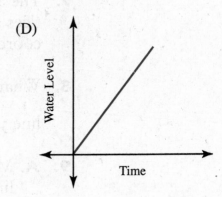

13. Do the arms of the parabola described by $x^2 + 3 + y = 6x - 1$ curve up or curve down?

14. Below are depictions of two relations.

   A. Write the set of ordered pairs for each relation.

   B. Determine if each relation is a function. If it is a function, write the linear equation that describes the function. If the relation is not a function, explain why it is not.

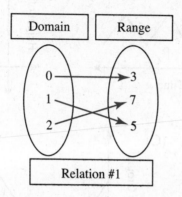

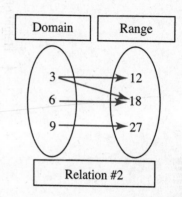

15. Graph the inequality $3y - 1 > 5 + 2x$.

16. Solve the system of equations to find two numbers whose sum is 13 and difference is 5.

17. Find the slope of the line that passes through the points $(2, -5)$ and $(6, -4)$.

18. A little girl was swinging at a steady rate on a swing in the playground. Which graph depicts the distance of the swing off the ground over time?

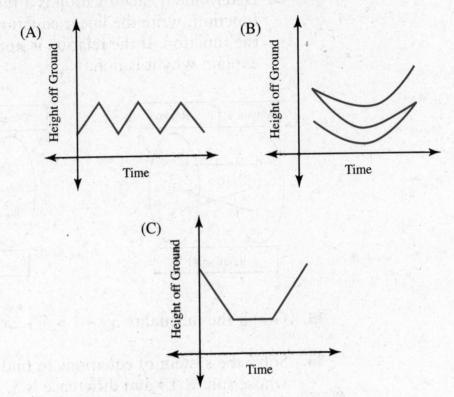

(A)

(B)

(C)

19. Solve the system of equations graphically:
$x + 3 = y$ and $2x = y + 7$.

20. During a thunderstorm, Forestville's weather station recorded that rain fell at $\frac{1}{2}$ inch per hour from 6 A.M. to 11 A.M. and then stopped for 3 hours. Then it began raining at a rate of $\frac{1}{4}$ inch per hour until 6 P.M. Draw a graph that shows the amount of rain that fell over this time period.

Chapter 9

Transformations

A *transformation* describes the relationship between a figure in a plane and its preimage. *Translations, reflections, rotations,* and *dilations* are types of transformations.

TRANSLATION

Triangle *ABC* was translated in the plane by shifting every point in the triangle 7 units to the left. A translation *slides* the figure to a new location on the plane. The image has the same size and shape as the original figure. This means that under a translation, the properties of the figure are preserved. The lengths of the sides are congruent, and the measures of the corresponding angles are equal.

$$A(0, 3) \Rightarrow A'(7, 3)$$
$$B(2, 7) \Rightarrow B'(9, 7)$$
$$C(4, 3) \Rightarrow C'(11, 3)$$

For any point *P* in $\triangle ABC$, $P(x, y) \Rightarrow P'(x - 7, y)$ represents the horizontal change of 7 units to the left.

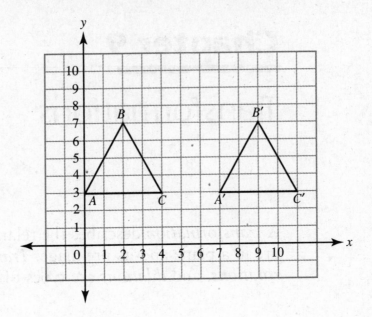

After a translation, the coordinates of parallelogram *ABCD* and its image are as follows.

$$A(7, 7) \Rightarrow A'(2, 3)$$
$$B(8, 10) \Rightarrow B'(3, 6)$$
$$C(9, 7) \Rightarrow C'(4, 3)$$
$$D(8, 4) \Rightarrow D'(3, 0)$$

Parallelogram *ABCD* was shifted 5 units horizontally left and 4 units vertically down. The translation relocated the points of the original figure $P(x, y)$ to its image $P'(x - 5, y - 4)$.

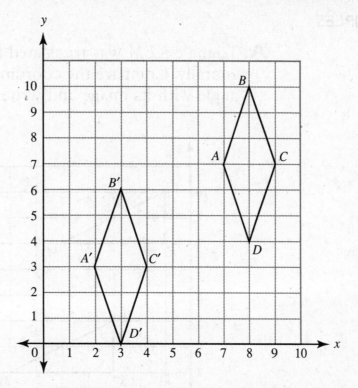

We can generalize the rule for a translation of any figure. Let *a* be the horizontal change (positive shift right and negative shift left), and let *b* be the vertical change (positive shift up and negative shift down).

$$P(x, y) \Rightarrow P'(x + a, y + b)$$

EXAMPLES

A. Triangle *KLM* was translated both vertically and horizontally. Compare the coordinates of the original triangle with its image and write the translation rule.

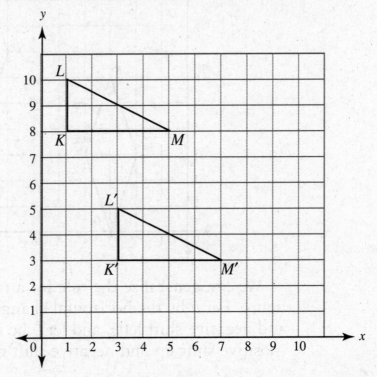

The translation shifted the original figure 2 units horizontally to the right and 5 units vertically down.

$$K(1, 8) \Rightarrow K'(3, 3)$$
$$L(1, 10) \Rightarrow L'(3, 5)$$
$$M(5, 8) \Rightarrow M'(7, 3)$$

Thus, $P(x, y) \Rightarrow P'(x + 2, y - 5)$

B. Find the coordinates of the image points of a triangle under a translation of $P'(x + 4, y + 1)$ for $A(0, 0)$, $B(1, 3)$, and $C(2, 1)$. Graph the image and label the vertices.

Use the translation rule.

$$A(0, 0) \Rightarrow (0 + 4, 0 + 1) \Rightarrow A'(4, 1)$$
$$B(1, 3) \Rightarrow (1 + 4, 3 + 1) \Rightarrow B'(5, 4)$$
$$C(2, 1) \Rightarrow (2 + 4, 1 + 1) \Rightarrow C'(6, 2)$$

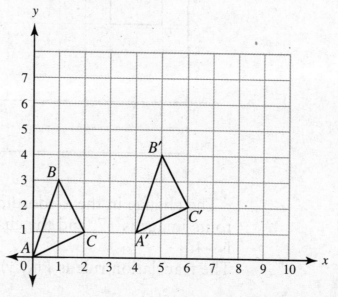

C. Rectangle $ABCD$ has been translated. Using the coordinates, write the rule for the translation if

$$
\begin{array}{lcl}
A(1, 7) & \Rightarrow & A'(8, 3) \\
B(3, 7) & \Rightarrow & B'(10, 3) \\
C(3, 3) & \Rightarrow & C'(10, -1) \\
D(1, 3) & \Rightarrow & D'(8, -1)
\end{array}
$$

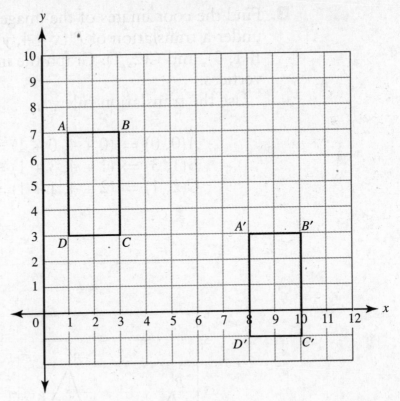

The change in the x-coordinates of the original point to its image is +7, and the change in the y-coordinates is −4.

The translation rule is $P(x, y) \Rightarrow P'(x + 7, y - 4)$.

SYMMETRY

The letters H, T, M, Y, and A all have one vertical line of symmetry. Draw a vertical line through the center of each of the letters. The shape on one side of the vertical line is a mirror image of the shape on the other side.

The letters that have one horizontal line of symmetry are B, C, D, and E.

The letters H, I, and X have two lines of symmetry.

A circle has an infinite number of lines of symmetry, with only four shown here.

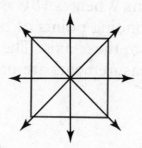

EXAMPLES

A. How many lines of symmetry does a square have?
A square has four lines of symmetry—one vertical, one horizontal, and two diagonal.

B. Does a rectangle also have four lines of symmetry like a square?

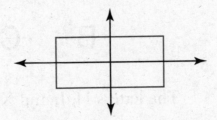

No. A rectangle has only a horizontal and a vertical line of symmetry.

The diagonals of a rectangle are not lines of symmetry. The shaded region is half of a rectangle. When this region is reflected over the line of symmetry, the dotted-line figure results.

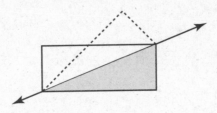

REFLECTION

In a reflection, the figure is *flipped* over the line of reflection. When △*ABC* is reflected over the *x*-axis, the corresponding points of both triangles are the same distance from the *x*-axis. The lengths of the sides and the angle measurements are preserved under a reflection.

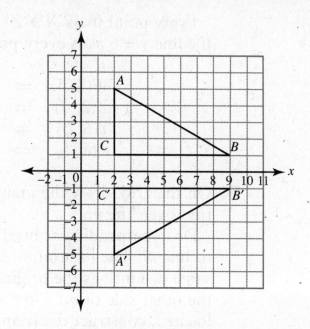

EXAMPLES

A. Reflect trapezoid WXYZ over the line $y = 6$.

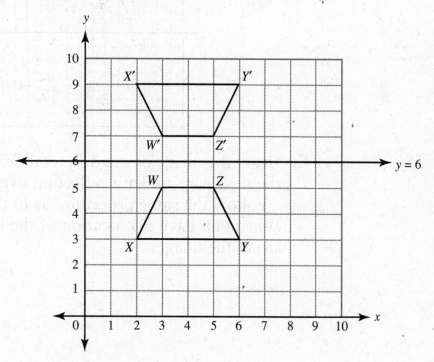

Every point in $W'X'Y'Z'$ is the same distance from the line $y = 6$ as is every point in $WXYZ$.

$$W(3, 5) \Rightarrow W'(3, 7)$$
$$X(2, 3) \Rightarrow X'(2, 9)$$
$$Y(6, 3) \Rightarrow Y'(6, 9)$$
$$Z(5, 5) \Rightarrow Z'(5, 7)$$

B. On the grid, draw the image of $\triangle KPG$ under a reflection over line m.

Drop perpendicular lines from points P, G, and K to line m. The location of each image point is the same distance as the original point from m but on the other side of m. After all three new points are located, construct the triangle.

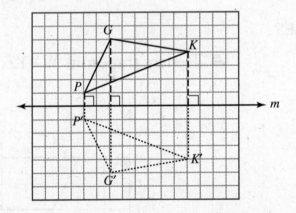

C. Draw the image of quadrilateral $WXYZ$ on the grid on the next page under a reflection over line l.

Follow the same procedure as in the above example. When you have the location of the image points, construct the image.

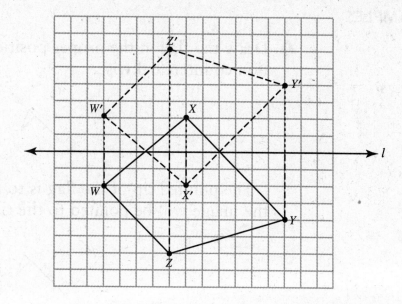

ROTATION

A *rotation* turns a figure about a fixed point in the plane. The letters N, S, and Z have rotational symmetry. The fixed point is the center of each letter, and the image is congruent with the original figure after a turn of 180°. The properties of the figures are preserved under a rotation.

N S Z

Another example of rotation symmetry uses the picture cards in a standard deck of cards. For example, place your finger in the center of the queen of diamonds. Turn the card 180°. The image you see now is the same as it was before you turned the card upside down.

EXAMPLES

A. Draw the flag in the proper position after a 90° rotation counterclockwise.

The pointed tip of the flag is to the right. The flag in the image will be pointed to the right also.

B. The hands of a clock now show 3:20. What time was it before the minute hand had rotated 72°?

There are 60 minutes in 1 hour and 360° in a circle. So each minute on a clock covers 6°. Divide 72° by 6° and get 12, which is the number of minutes the minute hand traveled over 72°. So 12 minutes before 3:20 was 3:08.

C. Rotate the rectangle *PQSR* 90° clockwise about the origin, and find the coordinates of the vertices of the image.

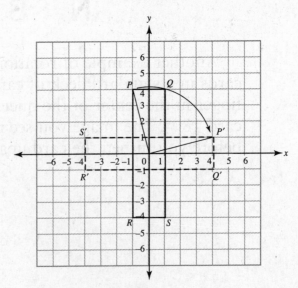

Think about this rotation as if you have your finger on the origin and turn the rectangle a quarter-turn to the right.

$$
\begin{aligned}
P(-1, 4) &\Rightarrow P'(4, 1) \\
Q(1, 4) &\Rightarrow Q'(4, -1) \\
R(-1, -4) &\Rightarrow R'(-4, -1) \\
S(1, -4) &\Rightarrow S'(-4, 1)
\end{aligned}
$$

D. Rotate $\triangle GDE$ 180° counterclockwise about G. List the coordinates of the new triangle.

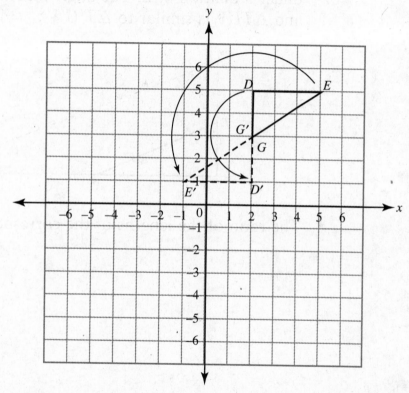

This is a half-turn counterclockwise about G. The coordinates of G and G' are the same, (2, 3). D(2, 5) $\Rightarrow$ D'(2, 1), and E(5, 5) $\Rightarrow$ E'(-1, 1).

DILATION

Images under a translation, reflection, or rotation are congruent with the original figure. The lengths of the sides and the measures of the angles are preserved. This is not so in a dilation. Just as the pupil of your eye shrinks or enlarges to adjust to the light, a geometric figure is shrunk or enlarged under a *dilation*. The angle measurement is preserved, while the length of each side is increased or decreased by a factor determined by the dilation.

Below, the lengths of the sides of $\triangle TUV$ are doubled under a dilation of 2. The angle measurement is preserved, and $\triangle TUV$ is similar to $\triangle T'U'V'$.

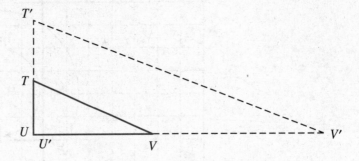

The ratio of the lengths of the corresponding sides is 1:2.

EXAMPLES

A. On the grid below, draw the image of $\triangle ABC$ after a dilation of 3. Label the vertices of the image $\triangle A'B'C'$.

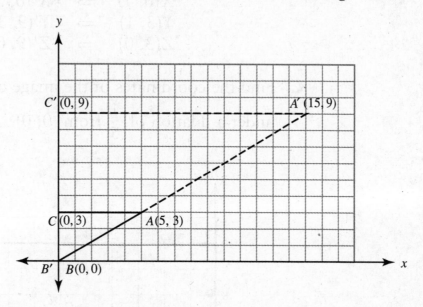

B. Find the factor by which rectangle $WXYZ$ was dilated to $W'X'Y'Z'$.

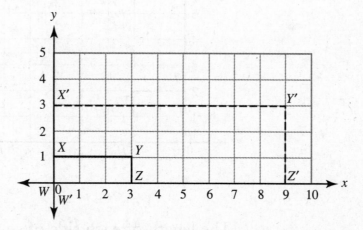

Find the lengths of the corresponding sides to determine their ratio.

$$\frac{WX}{W'X'} = \frac{1}{3}, \ \frac{XY}{X'Y'} = \frac{3}{9} = \frac{1}{3}, \ \frac{WZ}{W'Z'} = \frac{3}{9} = \frac{1}{3}, \text{ and } \frac{YZ}{Y'Z'} = \frac{1}{3}.$$

So this is a dilation of 3.

$$
\begin{array}{lcl}
W(0, 0) & \Rightarrow & W'(0, 0) \\
X(0, 1) & \Rightarrow & X'(0, 3) \\
Y(3, 1) & \Rightarrow & Y'(9, 3) \\
Z(3, 0) & \Rightarrow & Z'(9, 0)
\end{array}
$$

C. Find the coordinates of the image of square *JKLM* under a dilation of $\frac{1}{5}$ from (0, 0).

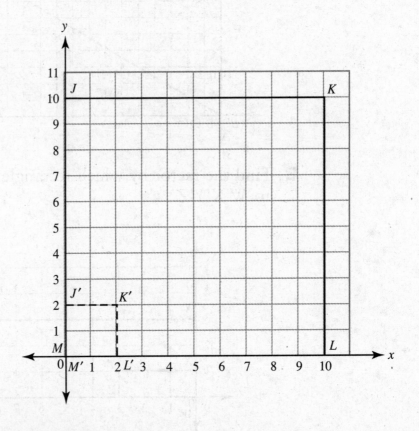

The length of each side of square *JKLM* is 10. A dilation of $\frac{1}{5}$ on *JKLM* reduces the length of each side to 2, where the coordinates of M and M' are (0, 0).

$$
\begin{array}{lcl}
J(0, 10) & \Rightarrow & J'(0, 2) \\
K(10, 10) & \Rightarrow & K'(2, 2) \\
L(10, 0) & \Rightarrow & L'(2, 0)
\end{array}
$$

TEST YOUR SKILLS   (See page 183 for answers.)

1. Which of these figures has rotational symmetry of 90°?

(A)   (C)

(B)   (D)

2. Triangles *B*, *C*, *D*, and *E* are transformations of △*A*. Which of the statements is false?
(A) △*B* is a dilation.
(B) △*C* is a translation.
(C) △*D* is a 90° rotation.
(D) △*E* is a reflection in the *y*-axis.

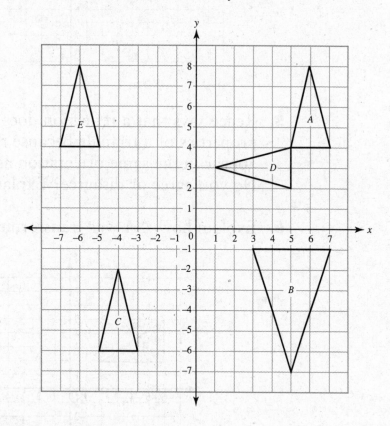

3. Point *S* was translated by the rule $(x + 6, y - 2)$ to its new location at $(3, -5)$. Which is the original location of Point *S*?
(A) (9, 1)   (B) (6, −2)   (C) (−1, −11)   (D) (−3, −3)

4. Reflect the quadrilateral over the *y*-axis, and then translate this image $A'B'C'D'$ down 6 units. List the coordinates of the new vertices of $A''B''C''D''$.

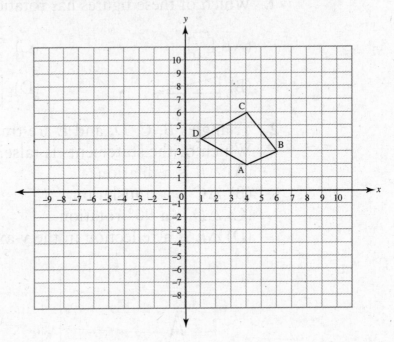

5. Renee says that a reflection does not preserve the properties of a triangle because the image does not appear in the same orientation as the original triangle. Do you agree or disagree? Explain your reasoning.

6. Explain how △*A* was transformed to its image △*B*.

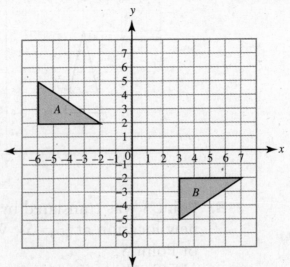

7. Suppose you are holding a starfish in your hand. You notice that all five of its arms are identical, so you know that this is an example of rotational symmetry. What is the minimum number of degrees you must rotate the starfish to show this rotational symmetry?

8. The point (–4, 3) is reflected over the y-axis. What are the coordinates of the image point?

9. Triangle ABC is translated down 2 units and then right 6 units. The coordinates are A(3, 2), B(3, 4), and C(5, 4).
   A. Write the formula for finding the coordinates of the image of these points.
   B. Find the coordinates of the image without graphing the figure.

10. Triangle XYZ has been translated left 8 units and up 3 units. The coordinates of the image are X′(3, –1), Y′(3, –3), and Z′(–1, –3). Find the coordinates of the *original* vertices X, Y, and Z.

11. Part of a figure is given in the diagrams below. Using both axes as lines of reflection, complete each figure.

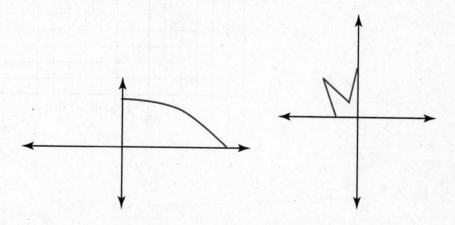

12. Find the coordinates of the new location of △JKL after a translation of (x – 4, y + 7), when the coordinates are J (–7, 4), K (–6, 7), and L (–4, 3).

13. A square, drawn on the coordinate plane with vertices at (0, 0), (0, 2), (2, 2), and (2, 0), is dilated by a factor of 3. The center point of the small square is (1, 1).

A. Find the coordinates of the larger square's center point.

B. Write the rule that translates the center of the small square to the center of the larger square.

14. Rectangle *KLMN* was reflected on the coordinate plane to *K'L'M'N'*. Write the equation for the line of reflection.

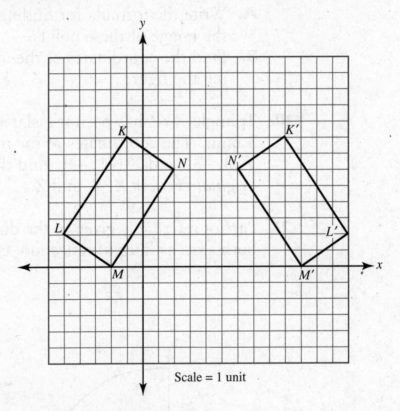

Scale = 1 unit

15. Triangle *ABC* was translated to its new location
△*A'B'C'* and then reflected over the line *y* = 3 to
become △*A"B"C"*.

   A. Draw △*A"B"C"* on the coordinate plane.
   B. Complete the table that shows the coordinates of
      the triangles.
   C. Write the rule for the translation of △*ABC* to
      △*A'B'C'*.

△**ABC**	△**A'B'C'**	△**A"B"C"**
A(–2, –2)	*A'*( )	*A"*( )
B(–5, –5)	*B'*( )	*B"*( )
C(–1, –4)	*C'*( )	*C"*( )

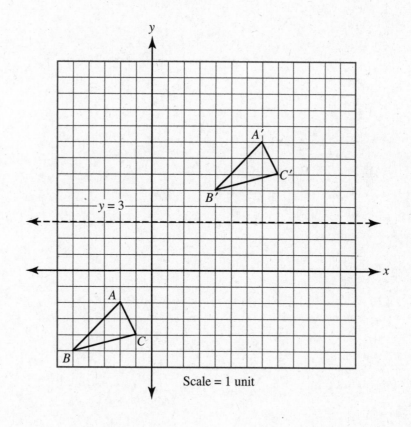

Scale = 1 unit

Chapter 10

Solutions to Practice Problems

The solutions presented here for the end-of-chapter problems are not the only possible solutions. You may know another way to solve the problem to arrive at the correct answer. Looking for different ways to solve the same problem helps improve your problem-solving skills.

Chapter 2: Numbers and Operations

1. (B) $4 + 3(15 - 3) \div 6 + 2^3 =$
$4 + 3 (12) \div 6 + 2^3 =$
$4 + 3 (12) \div 6 + 8 =$
$4 + 36 \div 6 + 8 =$
$4 + 6 + 8 = 18$

2. (B) According to the order of operations, the first operation you should use in this problem is division.

3. (D) $\dfrac{3}{22}$

Change to a mixed number.

$$7\frac{1}{3} = \frac{22}{3}$$

Then invert $\dfrac{22}{3}$ and get the reciprocal $\dfrac{3}{22}$.

4. (A) 12, 13
$12^2 = 144$ and $13^2 = 169$. Since $144 < 152 < 169$, 12 and 13 are the integers.

5. Substituting gives $4x^2 = 4(3^2) = 4(9) = 36$. Dave is correct. Work with exponents before multiplying.

6. (C) 16

    16: 1, 2, 4, 8, ⑯

    64: 1, 2, 4, 8, ⑯, 32, 64

    So 16 is the greatest common factor.

7. Divide 20 inches by 2½ inches to find the number of pieces.

$$20 \div 2\frac{1}{2} =$$

$$20 \div \frac{5}{2} = \frac{20}{1} \cdot \frac{2}{5} = \frac{40}{5} = 8$$

You can cut 8 pieces.

8. (D) $7.8 \cdot 10^5 \cdot 6.2 \cdot 10^3 =$

    $48.36 \cdot 10^8 =$

    $4.836 \cdot 10^9$

9.

$$\frac{4}{3}(6+9) =$$

$$\frac{4}{3}\left(\frac{6}{1}\right) + \frac{4}{3}\left(\frac{9}{1}\right) =$$

$$\frac{24}{3} + \frac{36}{3} = 8 + 12 = 20$$

10.

$$12 \cdot 3\frac{5}{6} =$$

$$\frac{12}{1} \cdot \frac{23}{6} = \frac{276}{6} = \frac{276 \div 6}{6 \div 6} = 46$$

Chapter 3: Patterns and Functions

1. (D) $12n - 5$

2. (A) The sum of eight times a number and five then divided by three.

3. (A) 0

Solve $x + 2 \leq -1$ for x.

$$x + 2 \leq -1$$
$$\underline{-2 = -2}$$
$$x \leq -3$$

The only choice *not* smaller than –3 is 0.

4. $\dfrac{6 - w^2}{5} = \dfrac{6 - (-9)^2}{5} = \dfrac{6 - 81}{5} = \dfrac{-75}{5} = -15$

5. A. Let h represent the number of hours the truck was used, and let c represent the cost.

$c = 50 + 10(h - 1)$

Subtract 1 from the number of hours used because that first hour's fee is covered under the $50.

B. $c = 50 + 10(h - 1)$
$c = 50 + 10(4 - 1)$
$c = 50 + 10(3)$
$c = 50 + 30$
$c = 80$

Therefore, the cost of renting the truck for 4 hours was $80.

6. $\dfrac{2,000}{6} \approx 333.\overline{3}$. So the lightbulb will last 333 days.

Since there are 365 days in a year and $333 < 365$, the bulb will not last 1 year.

7. A. $90s \leq 1,000$

B. $90s \leq 1,000$

$$\dfrac{90s}{90} \leq \dfrac{1,000}{90}$$
$$s \leq 11.\overline{1}$$

Thus, 11 students can get on the elevator in one trip.

8.

x	y
–2	0.5(–2) = –1
–1	0.5(–1) = –0.5
0	0.5(0) = 0
1	0.5(1) = 0.5
2	0.5(2) = 1

9. Let h be the number of hours worked.
$22.40 = 3h + 0.02(220)$
$22.4 = 3h + 4.4$
$18 = 3h$
$6 = h$
You worked for 6 hours.

10. The input value is multiplied by 4, and then 1 is added. The equation of the function is $y = 4x + 1$.

11. First find the weight of 1 box of paper.
10 reams at 4 pounds per ream ⇒ 1 box of paper weighs 40 pounds
If the cart can hold 700 pounds, then the number of boxes can be calculated by dividing 700 by 40.
$$\frac{700}{40} = 17.5$$
Therefore, the cart will hold 17 boxes of paper.

12. Let c be the cost of the letter and e be the cost of each ounce after the first ounce.
A. $c = 0.44 + 0.17(e - 1)$
B. $c = 0.44 + 0.17(e - 1)$
$c = 0.44 + 0.17(7 - 1)$
$c = 0.44 + 0.17(6)$
$c = 0.44 + 1.02$
$c = \$1.46$
A 7-ounce letter will cost \$1.46 to mail.

13. First, simplify the inequality.

$$6y + 5 \geq 2y + 7$$
$$\underline{-5 = \quad -5}$$
$$6y \quad \geq 2y + 2$$
$$\underline{-2y \quad = -2y}$$
$$4y \quad \geq 2$$

$$y \quad \geq \frac{2}{4} = \frac{1}{2}$$

Then graph the solution set on the number line.

14. The graph shows that the solution set contains numbers less than but not equal to −1. The inequality can be written as $x < -1$.

15. **A.** Let n be the number of pairs of socks Marie can buy.
$$1.98n \leq 25$$
B. $1.98n \leq 25$

$$n \leq \frac{25}{1.98} = 12.6262...$$

Marie can buy 12 pairs of socks.

Chapter 4: Ratio, Proportion, and Percent

1. (B) $\dfrac{2}{5} = \dfrac{4}{10} = \dfrac{40}{100} = 40\%$

2. (C) $\$10,000(0.03) = \$300$

3. (C) 7.2
$$0.006(1,200) = 7.2$$

4. Let x represent the original investment.
$$x + 0.04x = 15,080$$
$$1.04x = 15,080$$
$$x = 14,500$$

Micky originally invested \$14,500.

5. If 28% are left white, then 72% of the 50 squares should be shaded.
50(0.72) = 36 squares should be shaded.
You need to shade in 15 more squares.

6. Change both numbers to fractions or both to percents. Solving with fractions:

Kelly ate $10\% = \dfrac{1}{10}$ of the pizza.

Add together the girls' fractions. $\dfrac{1}{10} + \dfrac{3}{10} = \dfrac{4}{10} = \dfrac{2}{5}$

Together they ate $\dfrac{2}{5}$ of the pizza and left $\dfrac{3}{5}$ for the others.
Solving with percents:

Shelly ate $\dfrac{3}{10} = 30\%$ of the pizza.

Add 10% and 30% to get 40% of the pizza.
The girls ate 40% and left 60% of the pizza for the other guests.

7. You need 1 cup to make 1 qt. There are 4 cups in 1 qt, and 8 oz in 1 cup. Set up a proportion.

$$\dfrac{8 \text{ oz}}{32 \text{ oz}} = \dfrac{x}{24} \iff \dfrac{1}{4} = \dfrac{x}{24}$$

Solving for x, you get 6.

So 6 oz or $\dfrac{6}{8} = \dfrac{3}{4}$ cup of mix will make 24 oz of iced tea.

8. A. $8\% + \dfrac{1}{4}\% = 8\dfrac{1}{4}\% = 8.25\% = 0.0825$

B. To find the amount of tax, multiply 4 (the number of bookcases) by x (the cost of one bookcase) by the tax rate.
$t = 4x(0.0825)$

9. Find the number of students that have to be absent before the school can close, that is, 15% of 700.
0.15(700) = 105
The school cannot close because 105 students need to be absent for it to close and only 95 students are absent.

10. As a decimal number, 250% is equal to 2.5. Let x represent the jeweler's cost.
$2.5x = 2,085$
$x = 834$
The jeweler paid $834 for the necklace.

11. The first store:

Cost	$35.95
10% off	−3.60
Subtotal	$32.35
8% tax	+2.59
Total	$34.94

The second store:

Cost	$39.95
20% off	−7.99
Total	$31.96

Therefore, the second store has the better deal.

12. 2.4% of x is 21.
$0.024x = 21$
$$x = \frac{21}{0.024} = 875$$
So 875 students are at Greenfield.

13. The increase in cost of service is
$101.50 − $70 = $31.50
$$\frac{31.5}{70} = 0.45 = 45\%$$
The increase is 45%.

14.

$$prt = i$$

$$800(x)\left(\frac{1}{2}\right) = 4.8$$

$$400x = 4.8$$

$$x = \frac{4.8}{400} = 0.012 = 1.2\%$$

The interest rate is 1.2%

15. $\dfrac{7}{173} \approx 0.0404 \approx 4\%$

Thus, Ms. Harper will not include extra-credit questions on future tests.

Chapter 5: Measurement

1. (C) Since there are 36 in. in 1 yd, there are 18 in. in $\frac{1}{2}$ yd.

2. (D) Move the decimal point three places to the left.
   67 mm = 0.067 m

3. (D) 4 cups = 1 qt
       16 cups = 1 gal
       96 cups = 6 gal
   So, 96 8-ounce cups of iced tea are in 6 gallons.

4. Multiply \$55 by 0.83 to get €45.65.

5. 1 mi ≈ 1.6 km ≈ 1,600 m
   Divide 1,600 m by 4 to get 400 m in $\frac{1}{4}$ mi.

6. 1 gal = 4 qt, and 1 qt = 4 cups
   So 16 cups = 1 gal.
   1 gal = 16 cups ≈ 8 lb
   To find the weight of 1 cup, divide by 16.

   $$\frac{16 \text{ cups}}{16} = \frac{8 \text{ lb}}{16}$$

   1 cup weighs approximately $\frac{1}{2}$ lb or 8 oz.

7. 1 mi ≈ 1.6 km
   So 18 mi ≈ (18)(1.6) km

18 mi ≈ 28.8 km

The towns are approximately 28.8 km apart.

8. $F = \dfrac{9}{5}C + 32$

$F = \left(\dfrac{9}{5}\right)27 + 32$

$49 + 32 \approx 81°\text{F}$

Thus, 27°C ≈ 81°F.

9. (3 servings)(12 oz) = 36 oz per student

(36 oz)(24) = 864 oz

There are 32 oz in 1 qt, which means there are 128 oz in 1 gal.

864 oz ÷ 128 oz = 6.75 gal

If the parents bring 7 gal of orange drink, they will have enough.

10. 1 km ≈ $\dfrac{5}{8}$ mi, so 5 km ≈ $\dfrac{25}{8}$ or $3\dfrac{1}{8}$ mi. This race is too long for Paul.

11. (440 mi)$\left(\dfrac{8}{5}\right)$ ≈ 704 km. The campground is 270 km one way or 540 km round-trip. With a range of 704 km, the family will have plenty of gas for their vacation.

12. $60\left(\dfrac{5}{8}\right)$ = 37.5 or 38 mph

The car is traveling at 38 miles per hour.

13. 1 gal ≈ 3.8 liters; 10 gal ≈ 38 liters; 20 gal ≈ 76 liters. You need about 76 liters to fill a 20-gallon aquarium.

14. $98.6 = \dfrac{9}{5}C + 32$

$66.6 = \dfrac{9}{5}C$

$333 = 9C$

$C = 37$

98.6°F ≈ 37°C

A healthy person's body temperature is 37°C.

15. Set up a proportion.

$$\frac{€1.08}{\$1} = \frac{€23}{x}$$

$$1.08x = 23$$

$$x = \frac{23}{1.08}$$

$$x \approx 21.30$$

No, you need \$21.30 to purchase the watch, but have only \$20.

Chapter 6: Monomials and Polynomials

1. (A) Regroup the like terms together.
   $$7m + 9y + 12y - 4m =$$
   $$7m - 4m + 9y + 12y =$$
   $$3m + 21y$$

2. (C) To factor, you need to find two factors of -30 whose sum is -1. The only pair that works is -6 and $+5$.

3. (C) $(3x^2)^3 = (3^3)(x^2)^3 = 27x^6$

4. The area of the rectangle is the product of its length and width.
   $$(9x)(4x) = 36x^2 \text{ square units}$$

5. The formula for distance is $d = rt$, where r is the rate the plane is traveling and t is the time in hours. Multiply the binominals using FOIL.
   $$d = (y + 80)(4y - 3)$$
   $$d = 4y^2 - 3y + 320y - 240$$
   $$d = 4y^2 + 317y - 240$$

6. An equilateral triangle has three sides of equal length. The length of one side will be
   $$\frac{12x + 27}{3} = \frac{12x}{3} + \frac{27}{3} = 4x + 9 \text{ units}$$

7. Divide the polynomial by $5x^2$.

$$\frac{15x^4 + 20x^3 - 35x^2}{5x^2} =$$

$$\frac{15x^4}{5x^2} + \frac{20x^3}{5x^2} - \frac{35x^2}{5x^2} =$$

$$3x^2 + 4x - 7$$

8. First, find the length of the side of the square.

$$\frac{4p - 12}{4} = p - 3$$

The area of the square will be $(p - 3)^2$.
$(p - 3)^2 =$
$(p - 3)(p - 3) =$
$p^2 - 3p - 3p + 9 =$
$p^2 - 6p + 9$ units^2

9. Square the monomial.
$(6x^2y^3)^2 =$
$(6^2)(x^2)^2(y^3)^2 =$
$36x^4y^6$ square units

10. Notice that both d^2 and 81 are perfect squares. This binomial is the difference of two perfect squares, so the factors are $(d + 9)$ and $(d - 9)$.

11. Use the FOIL method to find the other factor. The first term in both factors is x. You are already given one factor. The other factor of +48 that when added to –8 gives you –14 is –6.
$(x - 8)(x - 6) =$
$x^2 - 14x + 48$
So the other factor is $(x - 6)$.

12. David has the incorrect factors. By multiplying his factors, you get $y^2 + y - 6$. The factors of –2 and +3 are correct. The correct binomial factors should be $(y + 6)$ and $(y - 1)$.

13. Factor $k^2 + 3k - 28$. Find two factors of –28 so that when they are added together, their sum is +3. The factors are +7 and –4. So the dimensions of the rectangle are $(k + 7)$ and $(k - 4)$.

14. Factor $x^2 - 11x + 24$. The numbers –8 and –3 have a product of +24 and a sum of –11. The factors are $(x - 8)$ and $(x - 3)$.

15. Find two factors of –18 that when added give you –3. The numbers –6 and +3 will work. The factors are $(x - 6)$ and $(x + 3)$.

Chapter 7: Geometry

1. (C) Point G is shared by both segments that form $\angle k$. Another name for $\angle k$ is $\angle BGM$.

2. (D) The sum of the interior angles of a triangle is 180°. Since $\angle Q$ is 90°, the sum of $\angle W$ and $\angle T$ must be 90°. So the angles are complementary.

3. (B) Angles 3 and 4 are supplementary. If $\angle 4 = 127°$, that means $\angle 3 = 53°$. Angles 3 and 6 are congruent because they are alternate interior angles. So $\angle 6 = 53°$. Angles 6 and 7 are congruent because they are vertical angles. So $\angle 7 = 53°$. The sum of angles 6 and 7 is 106°.

4. **A.** You can construct three equilateral triangles. One has sides of 2 cm, one has sides of 7 cm, and one has sides of 9 cm.

   B. There are six possible combinations of 2, 7, and 9, but only four triangles can be constructed. They can have sides of 2, 7, and 7 cm; 2, 9, and 9 cm; 7, 9, and 9 cm; and 9, 7, and 7 cm. You cannot construct triangles with sides of 2, 2, and 7 cm or 2, 2, and 9 cm because the sum of any two sides of a triangle *must* be greater than the third side.

   C. You cannot construct a scalene triangle with these lengths. The sum of 2 and 7 is 9, which would be the length of the third side. The sum of the lengths of any two sides of a triangle must be greater than the length of the third side.

5. *Reason 1:* The measurements of the corresponding angles are equal in similar triangles.

Reason 2: If ∠ACD is 180°, ∠ACE and ∠y are supplementary. The same is true for ∠ACE and ∠x, ∠BCD and ∠y, and ∠BCD and ∠x.

∠x + ∠ACE = 180° and ∠y + ∠ACE = 180°

∠x + ∠ACE = ∠y + ∠ACE

∠x = ∠y

Angles x and y are vertical angles, and vertical angles are always equal in measure.

6. Rotate the smaller triangle to see the corresponding sides more clearly.

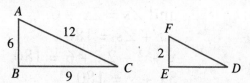

Pairs of corresponding sides in similar triangles have the same ratio.

$$\frac{AB}{FE} = \frac{BC}{ED} = \frac{AC}{FD}$$

To find $\overline{ED}$, set up a proportion.

$$\frac{AB}{FE} = \frac{BC}{ED}$$

Substitute the measure of the lengths for the sides.

$$\frac{6}{2} = \frac{9}{ED}$$

Cross-multiply and get the length of $\overline{ED}$ as 3 units. To find $\overline{FD}$, follow the same procedure using a proportion.

$$\frac{AB}{FE} = \frac{AC}{FD}$$

$$\frac{6}{2} = \frac{12}{FD}$$

Cross-multiply. $\overline{FD}$ = 4 units.

7. (A) Vertical angles are congruent. Choice B is false because trapezoids are not parallelograms. Trapezoids have only one pair of parallel sides. Choice C is false because the sum of two complementary angles is 90°. These two angles are supplementary because 86° + 94° = 180°.

8. **A.** Angles a and b are congruent because they are vertical angles. Angles b and d are congruent because they are alternate interior angles.

   B. The pairs of angles that are supplementary are c and d, b and c, and a and c.

9. Angles r and s are supplementary, where $\angle r = x + 3$ and $\angle s = 2x + 6$.
   $\angle r + \angle s = 180$
   $x + 3 + 2x + 6 = 180$
   $3x + 9 = 180$
   $3x = 171$
   $x = 57$
   Substitute the value of x to find $\angle s$.
   $\angle s = 2(57) + 6$
   $\angle s = 114 + 6$
   $\angle s = 120°$

10. Angles x, y, and z form a supplementary angle. Knowing that $\angle y = 90°$, we can say that angles x and z are complementary. We can solve the equation substituting m for the measure of $\angle x$.
    $m + 2m = 90$
    $3m = 90$
    $m = 30$
    So $\angle z = 2(30)$ or 60°.

11. Angles p, r, and s are congruent with each other ($\angle p$ and $\angle r$ are vertical angles, and $\angle p$ and $\angle s$ are corresponding angles). Let a represent the measure of one of these angles.
    $a + a + a = 180$
    $3a = 180$
    $a = 60°$

If $\angle s$ is 60°, then $\angle t$ is 120° because $\angle s$ and $\angle t$ are supplementary.

12. The pair of angles b and X and the pair c and Z are vertical angles. Vertical angles are congruent. The sum of the interior angles of a triangle is 180°.
$$\angle X + \angle Y + \angle Z = 180°$$
$$35° + \angle Y + 80° = 180°$$
$$\angle Y + 115° = 180°$$
$$\angle Y = 65°$$

13. Angles 3 and 7 are congruent, and angles 3 and 4 are supplementary. The sum of angles 3 and 4 is 180°.
$$3x + 2 + x - 10 = 180°$$
$$4x - 8 = 180°$$
$$4x = 188°$$
$$x = 47°$$
$$\angle 3 = \angle 7 = x - 10 = 47 - 10 = 37$$
Thus, $\angle 3 = 37°$

14. The base angles in an isosceles triangle are congruent. That means that $\angle a \cong \angle c$. The sum of the interior angles of a triangle is 180°. This leaves 90° for the sum of angles a and c since $\angle B$ is a right angle. So angles a and c each measure 45°.

    If $\angle a$ is 45°, then $\angle x$ is 135° because they are supplementary angles.

15. Angles a and b are supplementary angles. Their sum is 180°.
$$3x - 5 + 10x + 3 = 180$$
$$13x - 2 = 180$$
$$13x = 182$$
$$x = 14$$
$$\angle a = 3(14) - 5 = 42 - 5 = 37°$$
$$\angle b = 10(14) + 3 = 140 + 3 = 143°$$

Chapter 8: Graphing

1. **(B)** The equation of the line is written in slope-intercept form. The y-intercept, b, is 8. The coordinates of this point are $(0, 8)$.

2. **(B)** Substitute the coordinates into the equation. The only point that makes the statement true is $(-4, 4)$.

3. **(C)** Since the line is dashed and the shading is above the line, the inequality will begin with $y >$. The slope of the line rises 6 and runs 2. In other words, the slope is 3. The y-intercept is -1.

4. **(D)** Substitute each set of coordinates into the equation. The only point that does not lie on the line is $(6, 4)$.

$$y = \frac{1}{2}x - 1$$
$$6 \overset{?}{=} \frac{1}{2}(4) - 1$$
$$6 \overset{?}{=} 2 - 1$$
$$6 \neq 1$$

5. **(D)**
$$7x - y = 4$$
$$7x - 4 - y = 0$$
$$7x - 4 = y$$
$$y = 7x - 4$$

6. Use the slope-intercept form, $y = mx + b$. The slope, m, is 4. The y-intercept, b, is 7. The equation is $y = 4x + 7$.

7. The only possible location for the fourth vertex of a rectangle is (4, –4).

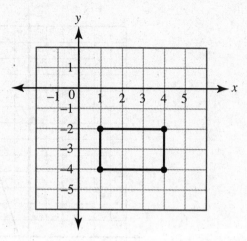

8. The slope of $y = \dfrac{-1}{2}x + 8$ is $-\dfrac{1}{2}$. The slope of the perpendicular line is the negative reciprocal, which is 2.

9. A. You can choose any values for x. This table shows one possible set of points.

x	$y = 3x - 2$	Point
–2	$y = 3(-2) - 2$ $y = -6 - 2$ $y = -8$	(–2, –8)
–1	$y = 3(-1) - 2$ $y = -3 - 2$ $y = -5$	(–1, –5)
0	$y = 3(0) - 2$ $y = -2$	(0, –2)
1	$y = 3(1) - 2$ $y = 3 - 2$ $y = 1$	(1, 1)
2	$y = 3(2) - 2$ $y = 6 - 2$ $y = 4$	(2, 4)

B. Plot (0, –2) and then count off the slope of 3.

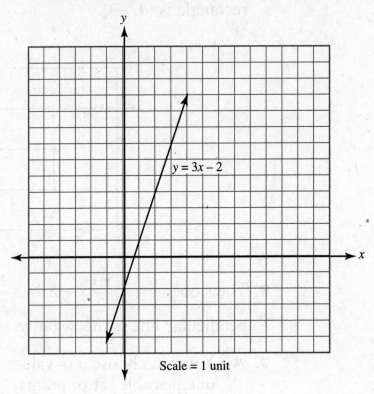

Scale = 1 unit

10. When the given points are connected, the line segment is not parallel to either axis. This segment must be the hypotenuse. Since the area of this right isosceles triangle is 18 units², this triangle must be half of a square because the triangle is isosceles. The figure is a square that has sides of length 6. The hypotenuse of this triangle is the diagonal of the square.

There are two possible answers to this problem. The third vertex can be (–1, 3) or (5, –3).

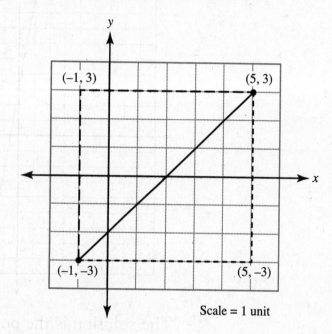

Scale = 1 unit

11. Rewrite the linear equations in slope-intercept form. Then graph the lines.

$$2y - 6 = x$$
$$2y = x + 6 \qquad \text{and} \qquad$$
$$y = \frac{1}{2}x + 3$$

$$6x + y = 3$$
$$y = -6x + 3$$

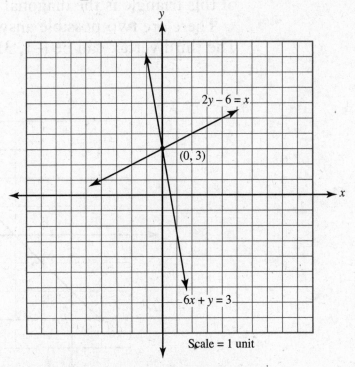

The solution is the point (0, 3).

12. (D) In graph A, the line shows the water level rising quickly and then slowly leveling off. This does not show a steady rate. Graph B shows a slow start and then the water level rising very quickly. Again, this is not a steady rate. Graph C shows a steady rate for a while, and then the rate slows and keeps steady. Graph D shows a steady rate of fill over the entire period of time.

13. Rewrite the quadratic equation in standard form.

$$x^2 + 3 + y = 6x - 1$$
$$x^2 + y = 6x - 4$$
$$y = -x^2 + 6x - 4$$

Since the sign on the x^2 term is negative, the arms will curve down.

14. A. $R_1 = \{(0, 3), (1, 5), (2, 7)\}$
$R_2 = \{(3, 12), (3, 18), (6, 18), (9, 27)\}$

B. The first relation, R_1, is a function since each element in the domain is paired with exactly one element in the range. The function is $y = 2x + 3$.

   The second relation, R_2, is not a function. The element 3 in the domain is paired with both 12 and 18.

15. Rewrite the inequality in slope-intercept form.

$$3y - 1 > 5 + 2x$$
$$3y > 6 + 2x$$
$$3y > 2x + 6$$
$$y > \frac{2}{3}x + 2$$

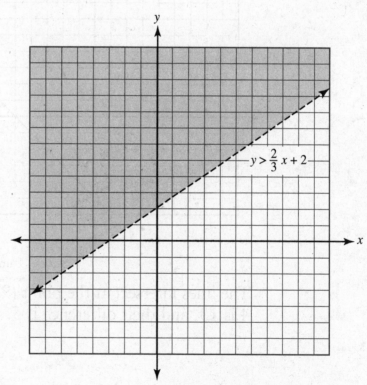

Scale = 1 unit

16. Write the two equations in slope-intercept form. Let x be the first number and y be the second number.

$$x + y = 13$$
$$y = -x + 13$$

and

$$x - y = 5$$
$$-y = -x + 5$$
$$y = x - 5$$

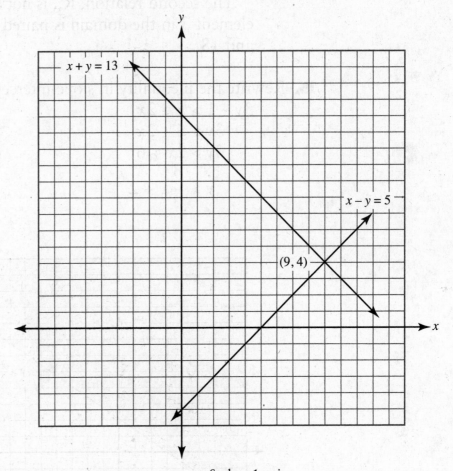

Scale = 1 unit

The lines intersect at the point (9, 4). The sum of 9 and 4 is 13, and their difference is 5.

17. Use the equation for slope. Let $(2, -5)$ be point 1 and $(6, -4)$ be point 2.

$$m = \frac{y_2 - y_1}{x_2 - x_1}$$

$$m = \frac{-4 - (-5)}{6 - 2} = \frac{1}{4}$$

The slope of the line is $\frac{1}{4}$.

18. (A) The little girl was swinging at a steady rate. So the maximum height of the swing off the ground is the same whether she has just gone forward or backward. The low points of the **V** shape show her height when closest to the ground.

19. Rewrite the equations in slope-intercept form and then graph.

$$x + 3 = y \qquad \text{and} \qquad \begin{aligned} 2x &= y + 7 \\ 2x - 7 &= y \\ y &= x + 3 \end{aligned} \qquad \begin{aligned} 2x &= y + 7 \\ 2x - 7 &= y \\ y &= 2x - 7 \end{aligned}$$

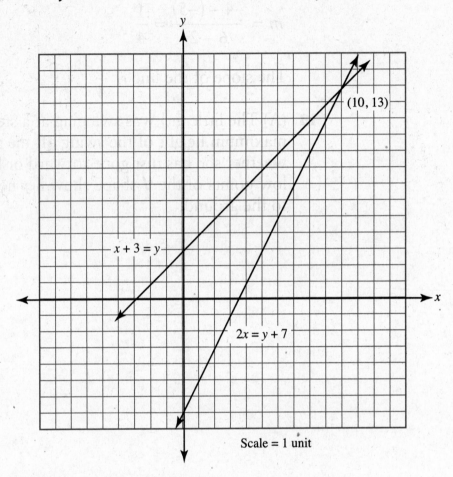

Scale = 1 unit

The solution is (10, 13).

20. Draw a grid with the horizontal axis labeled as time in hours and the vertical axis labeled as the rainfall amount in inches. Draw the line as you read the problem. When the rain stopped for 3 hours, remember that time did not. You must have a horizontal line that shows that no rain fell, but that time still passed.

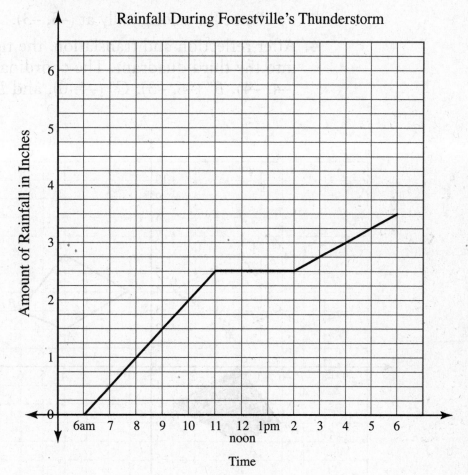

Chapter 9: Transformations

1. (A) The image of this figure after a quarter-turn is congruent with the original figure.

2. (B) Triangle B is not a dilation because the ratios of the base to the height of $\triangle A$ to $\triangle B$ are not the same.

$$\frac{2}{4} \neq \frac{4}{6}$$

3. (D) Use the rule to find the original location. Here, (x, y) are the coordinates of the original location, while $(3, -5)$ are the coordinates of the new location.

$$x + 6 = 3 \qquad \text{and} \qquad y - 2 = -5$$
$$x = -3 \qquad \text{and} \qquad y = -3$$

Point S was originally at $(-3, -3)$.

4. After reflection and translation, the figure is translated into the third quadrant. The coordinates are A'' $(-4, -4)$, B'' $(-6, -3)$, C'' $(-4, 0)$, and D'' $(-1, 2)$.

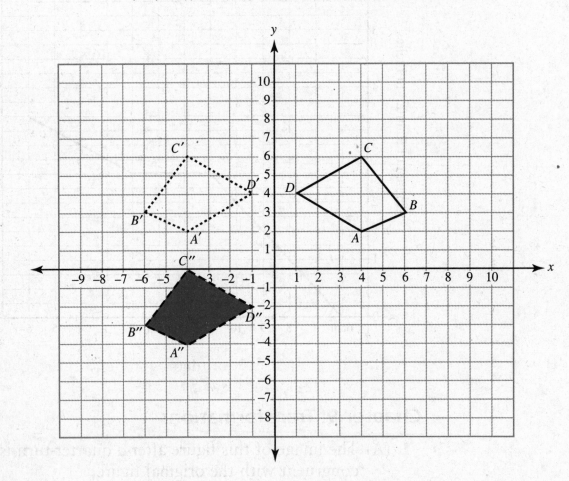

5. Reflections do preserve the properties of triangles. The corresponding sides are congruent, and the corresponding angles have the same measure as the original triangle. Orientation does not matter.

6. One way this can be accomplished is to reflect △A over the x-axis to become △A-1 and then translate it 9 units to the right to △B. You could also translate △A to △B-1 and then reflect △B-1 over the x-axis to △B.

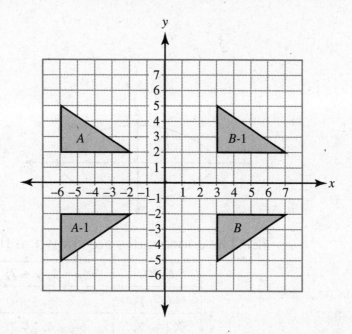

7. There are 360° in a circle, and one-fifth of 360° is 72°. You would have to turn the starfish 72°.

8. When reflected over the y-axis, the x-coordinate changes and the y-coordinate does not. The coordinate of the image point is (4, 3).

9. A. The translation changes the x-coordinate by +6, and the y-coordinate by −2. The formula is $(x + 6, y − 2)$.

B. $P(x, y) \Rightarrow P'(x + 6, y − 2)$
$A(3, 2) \Rightarrow A'(9, 0)$
$B(3, 4) \Rightarrow B'(9, 2)$
$C(5, 4) \Rightarrow C'(11, 2)$

10. The original triangle was translated to the left 8 units and up 3 units, $P'(x - 8, y + 3)$. To get the original coordinates, translate the image to the right 8 units and down 3 units. The vertices of the original triangle are found by the formula $P(x + 8, y - 3)$. So $X(11, -4)$, $Y(11, -6)$, and $Z(7, -6)$ are the original vertices.

11. Here are the figures.

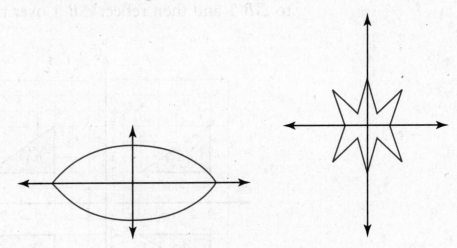

12. The coordinates are shown in the table below.

△JKL	(x - 4, y + 7)	△J'K'L'
J (-7, 4)	(-7 - 4, 4 + 7)	J' (-11, 11)
K (-6, 7)	(-6 - 4, 7 + 7)	K' (-10, 14)
L (-4, 3)	(-4 - 4, 3 + 7)	L' (-8, 10)

13. **A.** The center point of the larger square is at (3, 3).
B. To move the point (1, 1) to (3, 3), use the translation rule of $(x + 2, y + 2)$.

14. Divide the distance between N and N' and the distance between M and M'. The line of reflection has the equation $x = 4$.

15. A.

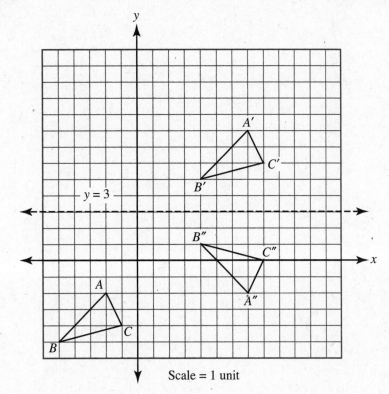

Scale = 1 unit

B. The coordinates are shown in the table below.

△*ABC*	△*A'B'C'*	△*A"B"C"*
A(−2, −2)	*A'*(7, 8)	*A"*(7, −2)
B(−5, −5)	*B'*(4, 5)	*B"*(4, 1)
C(−1, −4)	*C'*(8, 6)	*C"*(8, 0)

C. The translation rule is $(x + 9, y + 10)$.

Chapter 11

Sample Test 1

PART 1

Directions

You have 65 minutes to answer the following 42 multiple-choice questions. There is only one correct answer for each problem. Calculators are <u>NOT</u> allowed on this section of the test.

1. Which statement is true based on the diagram?

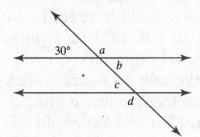

- **A.** $\angle b \cong \angle c$
- **B.** $\angle d = 30°$
- **C.** $\angle a$ and $\angle c$ are complementary.
- **D.** $\angle b$ and $\angle c$ are supplementary.

2. Simplify the expression below.

$$(11a^2b - ab + 5ab^2) - (3a^2b + 7ab^2)$$

- **A.** $14a^4b^4 - 10a^3b^3$
- **B.** $8a^2b - ab - 2ab^2$
- **C.** $8a^2b - ab + 12ab^2$
- **D.** $8a^2b - ab + 2ab^2$

3. Find x if $\angle A = x + 10$ and $\angle B = 5x - 10$.

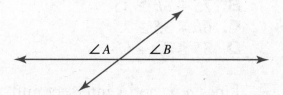

- **A.** 5
- **B.** 10
- **C.** 0
- **D.** 30

4. Multiply the two binomials below.

$$(3x + 1)(3x + 2)$$

- **A.** $9x^2 + 3$
- **B.** $3x^2 + 5x + 3$
- **C.** $9x^2 + 6x + 2$
- **D.** $9x^2 + 9x + 2$

5. The 16-gallon gas tank of a car was filled before a trip. After the trip, the gas gauge indicator shows the tank as one-quarter full. What percent of the gas in the tank was used on the trip?
A. 12%
B. 25%
C. 40%
D. 75%

6. Connie has $35. She makes $6 an hour babysitting. Which inequality shows the number of hours of babysitting needed for her to have at least $75?
A. $6h \geq 75 + 35$
B. $75 + 6h \geq 35$
C. $6h + 35 \geq 75$
D. $35 \leq 75 + 6h$

7. Lines q, r, and s intersect and form six angles. Which angles are not congruent?

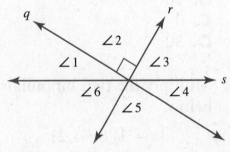

A. $\angle 2$ and $\angle 5$
B. $\angle 6$ and $\angle 3$
C. $\angle 5$ and $\angle 1$
D. $\angle 1$ and $\angle 4$

8. Which expression represents *3 times a number diminished by 7*?
A. $7(x - 3)$
B. $7 - 3x$
C. $3x - 7$
D. $3(7 - x)$

9. The school parking lot was three-eighths empty. What percentage of the lot was full?
A. 58%
B. 75%
C. 62.5%
D. 37.5%

10. Sally and Kira are selling Girl Scout cookies. Sally says to Kira, "If I sell this box and 3 more, I will have sold a total of 12 boxes." Kira replies, "If I sell 4 more boxes, I will have sold 16 boxes." Which statement below is correct?
A. Kira and Sally sold 15 boxes.
B. Sally sold 8 boxes.
C. Sally sold 3 more boxes than Kira.
D. Kira sold 7 boxes.

11. Which value of x will make the statement below true?

$$6(x - 7) - 5 = 19$$

A. 9
B. 11
C. 13
D. 15

12. Which expression is equivalent to $\dfrac{14x^5 + 35x^3}{7x^3}$?

A. $2x^2 + 5x^3$
B. $7x^3 (2x^2 + 5)$
C. $2x^2 + 5$
D. $2(x^2 + 7)$

13. Which expression describes $\angle A$ if $\angle B = 2x - 3$?

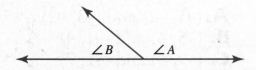

A. $180 - 2x + 3$
B. $90 + 2x + 3$
C. $180 - 2x - 3$
D. $90 + 2x - 3$

14. Which of the statements below is false?
A. The sum of the acute angles in a right triangle is 90°.
B. The sum of the interior angles in a quadrilateral is 360°.
C. Alternate interior angles are supplementary.
D. Alternate exterior angles are congruent.

15. Evaluate $5x^2$ when $x = 2$.
A. 20
B. 400
C. 100
D. 125

16. The hypotenuse of a right triangle is $\sqrt{29}$. If one leg measures 5, what is the length of the other leg?
A. 4
B. 3
C. 5
D. 2

17. Factor $x^2 + 5x + 6$.
A. $(x + 1)(x + 6)$
B. $(x + 2)(x + 3)$
C. $(x + 3)(x - 3)$
D. $(x + 5)(x + 1)$

18. Which expression is equivalent to $18x^3 + 12x^2$?

A. $6x(x^2 + 2x)$
B. $3x^2 + 2x^2$
C. $30x^5$
D. $6x^2(3x + 2)$

19. A video rental store sells previously viewed movies for $10 each. A sign above the display reads:

Buy	Save
2	$4
3	$8
4	$12
5	$16

Which expression best represents the savings based on the number of videos purchased?

A. $4(x - 1)$
B. $x + 4$
C. $4x$
D. $2x - 4$

20. After reflecting the shaded trapezoid over the y-axis and then reflecting the image over the x-axis, a new trapezoid appears in the second quadrant. Two vertices are plotted on the grid. What are the coordinates of the other two vertices of this new trapezoid?

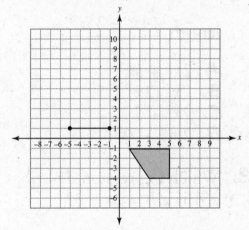

A. $(-5, -2)$ and $(-3, -2)$
B. $(-5, 4)$ and $(-1, 4)$
C. $(-9, 7)$ and $(-1, 7)$
D. $(-5, 4)$ and $(-3, 4)$

21. Which equation best represents the relation below?

x	−2	−1	0	1	2
y	−8	−3	2	7	12

A. $y = 4x$
B. $y = x - 2$
C. $y = 3x + 2$
D. $y = 5x + 2$

22. In the winter months, the classroom aquarium loses $\frac{1}{4}$ inch of water per week because of evaporation. Let f represent the height of the water in a full aquarium. Which equation best describes s, the amount of water in the aquarium after 6 weeks with only 1 inch of water added during that same period?

A. $f - 2 = \frac{1}{2}$

B. $f - \frac{6}{4} = s$

C. $\frac{3}{2} - 1 = s$

D. $f - \frac{1}{2} = s$

23. The top of a volleyball net is 6 feet off the ground. A line that is 10 feet long stabilizes the pole. How far from the base of the pole is the line?

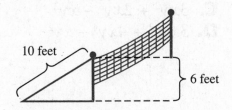

A. 16 ft
B. 136 ft
C. 6 ft
D. 8 ft

24. What is the product of $(5x^3 - 2x^2 + x - 3)$ and $2x$?
A. $5x^3 - 2x^2 + 3x - 3$
B. $7x^3 - 4x^2 + 2x - 6$
C. $10x^3 - 4x^2 + 2x - 3$
D. $10x^4 - 4x^3 + 2x^2 - 6x$

25. Given that the base angles in an isosceles triangle are congruent, find the measurement of $\angle BCD$.

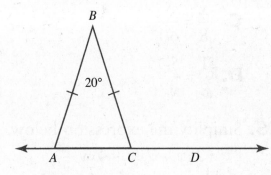

A. 120°
B. 80°
C. 100°
D. 110°

26. A motel can save 3% off its order of towels. If they usually pay $4 per towel, how much is saved when they purchase 3,000 towels?
A. $90
B. $360
C. $105
D. $144

27. The scale listed on the box containing a model ship is 1 inch to 8 feet. Which proportion can be used to find the actual length of the ship if the length of the model is 5 feet?

A. $\dfrac{1}{8} = \dfrac{x}{60}$

B. $\dfrac{1}{8} = \dfrac{60}{x}$

C. $\dfrac{x}{8} = \dfrac{5}{60}$

D. $\dfrac{1}{8} = \dfrac{5}{x}$

28. Simplify the expression below.

$$2xy + 7xy - 4xy$$

A. $13xy$
B. $5xy$
C. $5x^3y^3$
D. $5xy^3$

29. Which situation is best represented by $2m - 5$?
A. Paul has 2 less than 5 times the number of badges earned by Mitch, m.
B. Paul has 2 more than 5 times the number of badges earned by Mitch, m.
C. Paul has 5 less than twice the number of badges earned by Mitch, m.
D. Paul has 5 more than twice the number of badges earned by Mitch, m.

30. Solve for x: $4(3x - 1) = 8x + 12$
A. $x = 2$
B. $x = 4$
C. $x = -2$
D. $x = -4$

31. If $A = x^2 - 4x + 3$ and $R = -2x^2 + 5x + 3$, which expression represents $A - R$?
A. $3x^2 - 9x$
B. $-x^2 + 9x + 6$
C. $x^2 - x$
D. $-x^2 + 6$

32. Simplify the expression below.

$$(a^3b^2)(a^2b^4)$$

A. a^6b^8
B. a^5b^8
C. a^6b^6
D. a^5b^6

33. Find the quotient of the expression below.

$$\dfrac{12x^4y^2 + 8x^3y^4 - 4x^2y^2}{4x^2y}$$

A. $3x^2y + 2xy^3 - y$
B. $8x^2y + 4xy^3 - 3y$
C. $3xy^2 + 2xy^3 - xy^2$
D. $3x^2y^2 + 4xy^4 - xy^2$

34. Which inequality best represents this statement?

Five more than 3 times a number is greater than 6

A. $5x + 3 > 6$
B. $5 + x > 6$
C. $3x + 5 > 6$
D. $6 + 5x > 3$

35. Simplify the expression $(2a^2b^3)^3$.
A. $2a^6b^9$
B. $8a^5b^6$
C. $8a^6b^9$
D. $2a^5b^6$

36. Lines k, l, and m intersect at one point. Which pair of angles must be congruent?

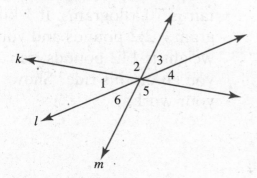

A. $\angle 2$ and $\angle 6$
B. $\angle 3$ and $\angle 4$
C. $\angle 1$ and $\angle 4$
D. $\angle 3$ and $\angle 5$

37. Two parallel lines are cut by a transversal. Which equation can be used to solve for x?

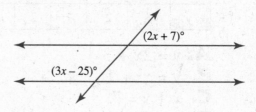

A. $3x - 25 + 2x + 7 = 90$
B. $3x - 25 + 2x + 7 = 180$
C. $2x - 25 = 3x + 7$
D. $x + 7 = 3x - 25$

38. Sue has a 6-gallon container. How many quarts of juice will this container hold?
A. 12
B. 24
C. 48
D. 64

39. Find the measure of $\angle x$.

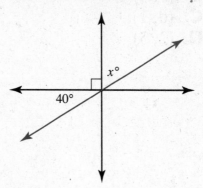

A. 50°
B. 40°
C. 130°
D. 140°

40. Which equation best represents the values in the table?

x	−2	−1	0	1	2
y	5	2	1	2	5

A. $y = 2x + 3$
B. $y = x^2 + 1$
C. $x + y = 7$
D. $y = 2x + 1$

41. Solve for x.

$$4(x + 3) + 2x - 7 = 3(x + 2) + 2(x - 1)$$

A. $x = -1$
B. $x = 3$
C. $x = -11$
D. $x = 8$

42. Which point does not lie on the line $y = 2x + 1$?
A. $(-1, -1)$
B. $(3, 4)$
C. $(0, 1)$
D. $(2, 5)$

PART 2

Directions

You have 70 minutes to answer the following extended-response questions. Be sure to explain your solutions clearly. Write legibly and neatly. You may use your calculator to solve these questions.

1. A. Write an equation to represent the statement below.

Six more than four times a number is 34.

B. Solve the equation.

2. The minimum weight for the Sling Shot ride at the county fair is 70 kilograms. If 1 kilogram ≈ 2.2 pounds and your weight is 112 pounds, can you go on this ride? Show your work.

3. An 80-foot flagpole casts a 30-foot shadow. Find the height of the tree if its shadow is 9 feet. Show your work.

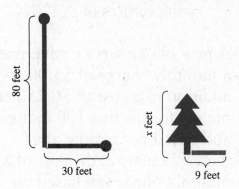

4. Two of the vertices of a right triangle are $A(4, 1)$ and $C(4, 9)$.

A. Plot and label these two points on the grid below.

B. Find the coordinates of B, which is in the second quadrant, so that the area of this triangle is 36 units². Explain how you arrived at your answer.

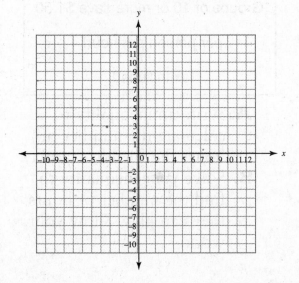

5. In the figure below, $\angle A = 7x$ and $\angle B = (4x + 26)°$. Find the measure of $\angle B$ in degrees.

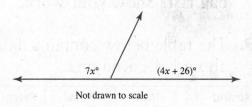

Not drawn to scale

6. Below is a table that follows a function rule.

x	y
−4	
0	1
2	
8	5
12	7

A. Write the equation of this function.

B. Fill in the missing numbers.

7. Lines a and b are parallel. Line t is a transversal.

A. If $\angle d = 50°$, find the measure of $\angle f$ in degrees. Show your work.

B. Which other angles are congruent to $\angle d$?

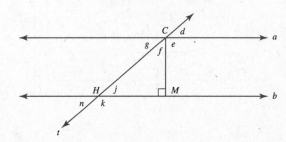

8. Kim's kitten eats 4 ounces of kitten food each day. How many days will a 10-pound bag last? Show your work.

9. The table below contains data about hot cocoa mix.

Brand	Total Weight	Total Price per Container	Price per Ounce
The Cocoa Hut	24 oz	$4.32	
Kid Cocoa	48 oz	$6.72	

 A. Calculate the price per ounce.
 B. Which brand is less expensive per ounce?

10. On the grid below, rotate $\triangle ABC$ 180° counterclockwise about the origin.
 A. Draw the image and label it $\triangle A'B'C'$.

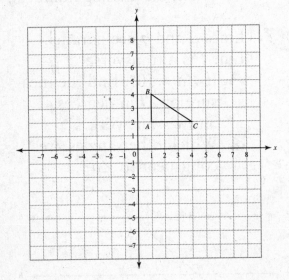

B. Translate $\triangle A'B'C'$ horizontally right 6 units and vertically down 2 units. What are the coordinates of the vertices of $\triangle A''B''C''$?

11. A new phone service advertises a monthly charge of $5.00 in addition to a cost of $0.25 per minute for the first 100 minutes and $0.10 per minute after that. Calculate c, the cost of a monthly phone bill based on 230 minutes of calling time. Show your work.

12. For his birthday, Adam took 11 of his friends to play laser tag. The sign below shows the rates. The group of 12 friends has $220.

RATES

*$5 per hour

*Groups of 10 or more save $1.50

per person per hour!

Tax included.

 A. Write an inequality to calculate the number of hours the friends can play.
 B. Solve your inequality to find the number of hours they can play. Show your work.

Answers to Sample Test 1

The solutions presented here for this sample test are not the only possible solutions. You may know another way to solve the problem and arrive at the correct answer. Looking for different ways to solve the same problem helps improve your problem-solving skills.

PART 1

1. A	10. B	19. A	28. B	37. B
2. B	11. B	20. D	29. C	38. B
3. D	12. C	21. D	30. B	39. A
4. D	13. A	22. D	31. A	40. B
5. D	14. C	23. D	32. D	41. A
6. C	15. A	24. D	33. A	42. B
7. C	16. D	25. C	34. C	
8. C	17. B	26. B	35. C	
9. C	18. D	27. B	36. C	

1. (A) $\angle b \cong \angle c$.

Choice B is false because $\angle d$ is obtuse. Choice C is false because $\angle a$ and $\angle c$ are not complementary but are supplementary. Choice D is false because $\angle b$ and $\angle c$ are not supplementary. They are both 30° and congruent.

2. (B) $8a^2b - ab - 2ab^2$

$(11a^2b - ab + 5ab^2) - (3a^2b + 7ab^2) =$
$11a^2b - ab + 5ab^2 - 3a^2b - 7ab^2 =$
$11a^2b - 3a^2b - ab + 5ab^2 - 7ab^2 =$
$8a^2b - ab - 2ab^2$

3. (D) 30

Angles A and B are supplementary, so $\angle A + \angle B = 180$.
$x + 10 + 5x - 10 = 180$
$x + 5x + 10 - 10 = 180$
$6x + 0 = 180$
$6x = 180$
$x = 30$

4. (D) $(3x + 1)(3x + 2)$

$(3x + 1)(3x + 2) =$
$9x^2 + 6x + 3x + 2 =$
$9x^2 + 9x + 2$

5. (D) 75%

If the tank is now one-quarter full, then three-quarters of the gas was used on the trip.

$\dfrac{3}{4} = 75\%$

6. (C) $6h + 35 \geq 75$

The money she now has ($35) plus the amount she makes at babysitting ($6h$) should be greater than or equal to $75.

7. (C) $\angle 5$ and $\angle 1$

The other three choices are vertical angles, which are congruent.

8. (C) $3x - 7$

The phrase *3 times a number* is represented by $3x$. It is *diminished by* 7 means it is made smaller by 7. This means 7 is subtracted from $3x$.

9. (C) The parking was three-eighths empty or five-eighths full.

$5 \div 8 = 0.625$ or 62.5% full.

10. (B) Sally sold 8 boxes.

From Kira's statement, you know she sold 12 boxes because $12 + 4 = 16$. Sally sold 8 boxes because *this box* $+ 3 + 8 = 12$.

11. (B) 11

$$6(x - 7) - 5 = 19$$
$$6(x - 7) - 5 + 5 = 19 + 5$$
$$6(x - 7) = 24$$
$$x - 7 = 4 \qquad \text{(after dividing both sides by 6)}$$
$$x = 11$$

12. (C) $2x^2 + 5$

$$\frac{14x^5 + 35x^3}{7x^3} =$$
$$\frac{14x^5}{7x^3} + \frac{35x^3}{7x^3} =$$
$$2x^2 + 5$$

13. (A) $180 - 2x + 3$

$\angle A = 180 - (2x - 3) = 180 - 2x + 3$

14. (C) Alternate interior angles are supplementary.

Since alternate interior angles are congruent, they are not supplementary.

15. (A) 20

$5x^2$ when $x = 2 \Rightarrow 5x^2 = 5(2^2) = 5(4) = 20$

The order of operations states that exponents should be simplified before multiplying.

16. (D) 2

$$a^2 + b^2 = c^2$$
$$5^2 + b^2 = \left(\sqrt{29}\right)^2$$
$$25 + b^2 = 29$$
$$b^2 = 4$$
$$b = 2$$

17. (B) $(x + 2)(x + 3)$

You need to find two factors of +6 that when added together give +5. The only factors that work are +2 and +3.

18. (D) $6x^2(3x + 2)$

The GCF of $18x^3 + 12x^2$ is $6x^2$.

$$\frac{18x^3}{6x^2} + \frac{12x^2}{6x^2} = 3x + 2$$

19. (A) $4(x - 1)$

Look for the pattern. Subtract 1 from the value in the *Buy* column and then multiply by 4.

20. (D) (–5, 4) and (–3, 4)

Reflect the shaded trapezoid into the third quadrant over the *y*-axis. Reflect this trapezoid over the *x*-axis. The coordinates of the two missing vertices are (–5, 4) and (–3, 4).

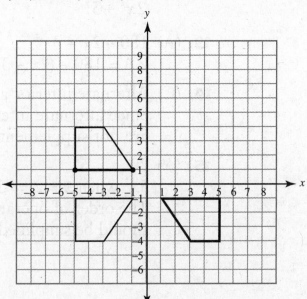

21. (D) $y = 5x + 2$

From $(0, 2)$, you can eliminate choices A and B. Test the values in the remaining two equations.

22. (D) $f - \dfrac{1}{2} = s$.

Over 6 weeks, the aquarium lost $\dfrac{1}{4}$ in. of water. The total loss was $\dfrac{1}{2}$ in. since 1 in. of water was added. Starting with a full tank, f, subtract $\dfrac{1}{2}$ in. of water to get s.

23. (D) 8 ft

6-8-10 is a multiple of the Pythagorean triple 3-4-5.

24. (D) $10x^4 - 4x^3 + 2x^2 - 6x$

Using the distributive property, multiply $2x$ by each term in the polynomial.
$(2x)(5x^3) = 10x^4$
$(2x)(-2x^2) = -4x^3$
$(2x)(x) = 2x^2$
$(2x)(-3) = -6x$

25. (C) $100°$

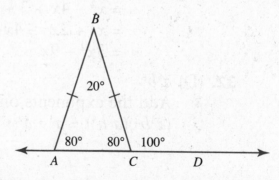

$\angle BAC$ and $\angle BCA$ are both $80°$. Since $\angle BCA$ and $\angle BCD$ are supplementary, $\angle BCD = 100°$.

26. (B) $360

$4(3,000) = $12,000

3% of $12,000 is $(0.03)(12,000) = $360

27. (B) $\dfrac{1}{8} = \dfrac{60}{x}$

Compare the model to the actual length. Convert the 5 feet to 60 inches.

28. (B) $5xy$

Since all three terms have the same variables, add the coefficients to get 5. Then add on the variables xy.

29. (C) Paul has 5 less than twice the number of badges earned by Mitch, m.

The minus sign on the 5 helps eliminate choices B and D. Since the number is being reduced by 5, eliminate choice A.

30. (B) $x = 4$

Simplify using your algebra skills.
$$4(3x - 1) = 8x + 12$$
$$12x - 4 = 8x + 12$$
$$4x - 4 = 12$$
$$4x = 16$$
$$x = 4$$

31. (A) $3x^2 - 9x$
$$A - R = (x^2 - 4x + 3) - (-2x^2 + 5x + 3)$$
$$= x^2 - 4x + 3 + 2x^2 - 5x - 3$$
$$= x^2 + 2x^2 - 4x - 5x + 3 - 3$$
$$= 3x^2 - 9x$$

32. (D) a^5b^6

Add the exponents on the same variable.
$$(a^3b^2)(a^2b^4) = a^3 \cdot a^2 \cdot b^2 \cdot b^4 = a^5b^6$$

33. (A) $3x^2y + 2xy^3 - y$

Simply divide each term in the numerator by the denominator.

$$\frac{12x^4y^2 + 8x^3y^4 - 4x^2y^2}{4x^2y} =$$

$$\frac{12x^4y^2}{4x^2y} + \frac{8x^3y^4}{4x^2y} - \frac{4x^2y^2}{4x^2y} =$$

$$3x^2y + 2xy^3 - y$$

34. (C) $3x + 5 > 6$

Five more ($+5$) than 3 times a number ($3x$) gives you $3x + 5$. Is greater than 6 (> 6) is shown in choice C.

35. (C) $8a^6b^9$

Do not forget to raise 2 to the power of 3, so choices A and D are eliminated. Rules for exponents state that when raising a power to a power, multiply the exponents.

36. (C) $\angle 1$ and $\angle 6$

Vertical angles are congruent. Choice C has the only pair of vertical angles.

37. (B) $3x - 25 + 2x + 7 = 180$

The two angles are supplementary. Their sum is $180°$.

38. (B) 24

There are 4 quarts in 1 gallon, or $6 \cdot 4$ or 24 quarts in 6 gallons.

39. (A) $50°$

The right angle $+ 40° + x° = 180°$ since they are supplementary angles. You can also think of this as $40° + x° = 90°$.

40. (B) $y = x^2 + 1$

Notice that the values for y are symmetrical. It should give you a hint that this is a parabola, or quadratic equation. You can also substitute the x-values in each equation.

41. (A) $x = -1$

Simplify using algebra.
$$4(x + 3) + 2x - 7 = 3(x + 2) + 2(x - 1)$$
$$4x + 12 + 2x - 7 = 3x + 6 + 2x - 2$$
$$6x + 5 = 5x + 4$$
$$x = -1$$

42. (B) $(3, 4)$

Substitute the values in the equation.
$$y = 2x + 1$$
$$4 \overset{?}{=} 2(3) + 1$$
$$4 \overset{?}{=} 6 + 1$$
$$4 \neq 7$$

PART 2

1. A. *Six more than four times a number is 34* is written as $6 + 4n = 43$ or $4n + 6 = 34$.

B. Solved:
$$4n + 6 = 34$$
$$4n + 6 - 6 = 34 - 6$$
$$4n = 28$$
$$n = 7$$

2. Solve using a proportion to find the number of kilograms that approximately equal 112 lb.

$$\frac{1\,\text{kg}}{2.2\,\text{lb}} \approx \frac{x\,\text{kg}}{112\,\text{lb}}$$

$$2.2x \approx 112$$

$$x \approx 50.9\,\text{kg}$$

No, you are not able to go on the ride.

3. Use a proportion to find the height of the tree.

$$\frac{x}{9} = \frac{80}{30}$$

$$30x = 720$$

$$x = 24$$

The tree is 24 ft tall.

4. A.

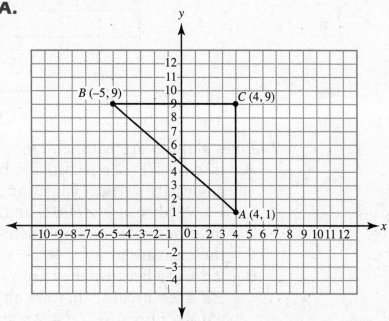

B. Points A and C are graphed on the grid. If the area of the triangle is 36 units², and the height, AC, of the triangle is 8 units, the base of the triangle has to be 9 units because the area of the triangle is found with the formula $A = \frac{1}{2}bh$. The point in the second quadrant that satisfies this requirement is $B(-5, 9)$.

5. Angles A and B are supplementary, so their sum is 180°.

$\angle A + \angle B = 180°$

$7x + 4x + 26 = 180°$

$11x + 26 = 180$

$11x = 154$

$x = 14$

$\angle B = 4x + 26 = 4(14) + 26 = 56 + 26 = 82$

Thus, $\angle B = 82°$.

6. A. Observing the pattern, y is obtained by taking half the value of x and adding 1. The equation is $y = \dfrac{x}{2} + 1$.

B.

x	y
–4	–1
0	1
2	2
8	5
12	7

The missing values of y are –1 and 2.

7. Lines a and b are parallel. Line t is a transversal.

A. If $\angle d = 50°$, $\angle g$ is also 50° because these two angles are vertical angles and are congruent. Angle $e = 90°$ because $\angle HMC$ and $\angle e$ are alternate interior angles and are therefore congruent. So the measure of $\angle f = 40°$.

B. $\angle d \cong \angle g$ because they are vertical angles. $\angle d \cong \angle n$ because they are alternate exterior angles. $\angle d \cong \angle j$ because $\angle d \cong \angle n$ and $\angle n \cong \angle j$ because $\angle n$ and $\angle j$ are vertical angles.

8. There are 16 oz in 1 lb, or 160 oz in the 10-lb bag. At 4 oz each day, the bag will last $\dfrac{160}{4}$ or 40 days.

9.

Brand	Total Weight	Total Price per Container	Price per Ounce
The Cocoa Hut	24 oz	$4.32	$0.18
Kid Cocoa	48 oz	$6.72	$0.14

A. Use your calculator. The price per ounce of The Cocoa Hut hot chocolate is $4.32 ÷ 24 or $0.18. Kid Cocoa is priced at $6.72 ÷ 48 or $0.14 per ounce.

B. Kid Cocoa is less expensive per ounce.

10. A. The image of △*ABC* is in the third quadrant.
    A′ is located at (–1, –2), *B′* is at (–1, –4), and *C′* is
    at (–4, –2).

   B. The vertices of △*A″B″C″* are *A″*(5, –4),
    B″(5, –6), and *C″*(2, –4).

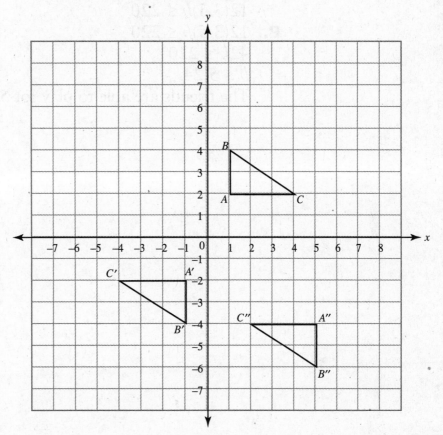

11. The monthly bill is calculated by adding the $5 charge
plus the cost for 230 minutes.
$$5 + (0.25)(100) + (0.10)(130) = c$$
$$5 + 25 + 13 = c$$
$$c = \$43$$

12. A. The rate per child is $3.50 ($5 − $1.50 = $3.50) per hour. Multiply the cost per hour by the number of children by the number of hours to get the total cost, which has to be equal to or less than $220.

$12(3.5)h \leq 220$

B. $12(3.5)h \leq 220$

$42h \leq 220$

$h \leq 5.24$

The friends are able to play for 5 hours.

Chapter 12

Sample Test 2

PART 1

Directions

You have 65 minutes to answer the following 42 multiple-choice questions. There is only one correct answer for each problem. Calculators are <u>NOT</u> allowed on this section of the test.

1. Which expression is a simplified form of $\dfrac{3x^3 - 9x^2}{3x^2}$?

  A. $3x^2(x - 3x)$
  B. $3x(x^2 - 3)$
  C. $3(x^3 - 3x)$
  D. $x - 3$

2. In the accompanying diagram, parallel lines r and s are cut by transversal m. Which angles are congruent?

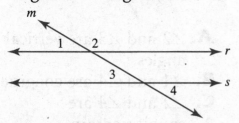

  A. $\angle 2$ and $\angle 3$
  B. $\angle 1$ and $\angle 4$
  C. $\angle 2$ and $\angle 4$
  D. $\angle 1$ and $\angle 2$

3. Evaluate the expression $2x^3$ when $x = 3$.

  A. 54
  B. 18
  C. 63
  D. 216

4. Simplify the expression below.

$9k^2m^3 + 2k^2m^3$

  A. $11k^2m^3$
  B. $11k^4m^6$
  C. $18k^2m^3$
  D. $18k^4m^9$

5. Kim's class is selling plants as a fund-raiser. The nursery charges $6 per plant, which the students will sell for $8. Which inequality can be used to find the number of plants that must be sold for the class to make at least $75?

A. $6x \geq 75$
B. $8 - 2x < 75$
C. $6x - 2 \geq 75$
D. $2x \geq 75$

6. Which are the binomial factors of $y^2 - 4y - 12$?

A. $(y - 8)$ and $(y + 4)$
B. $(y - 2)$ and $(y + 6)$
C. $(y + 2)$ and $(y - 6)$
D. $(y - 3)$ and $(y + 4)$

7. Angles WXY and YXZ are supplementary. Which expression represents the measurement of $\angle YXZ$?

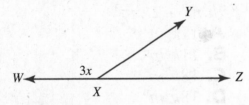

A. $90 + 3x$
B. $180 + 3x$
C. $90 - 3x$
D. $180 - 3x$

8. Which sentence is equivalent to the equation $3m + 4 = g$?

A. The number of guppies is 4 more than 3 times the number of mollies
B. The number of guppies is 8 less than the number of mollies.
C. The number of guppies is 3 more than 4 times the number of mollies.
D. The number of guppies is 4 less than 3 times the number of mollies.

9. Multiply the binomials.
$$(2x + 5)(3x - 4)$$

A. $5x + 1$
B. $6x^2 - 20$
C. $6x^2 + 7x - 20$
D. $6x^2 + 23x - 20$

10. Which statement is correct based on the diagram?

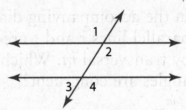

A. $\angle 2$ and $\angle 3$ are vertical angles.
B. $\angle 1$ and $\angle 4$ are congruent.
C. $\angle 2$ and $\angle 4$ are supplementary.
D. $\angle 1$ and $\angle 3$ are complementary.

11. A certain variety of tomato seed, s, has a 15% germination failure rate. Based on this information, how many seeds should be planted to produce 200 plants?

A. $200 = 1.15s$
B. $200 = s + 0.85$
C. $200 = 1.85s$
D. $200 = s + 185\%$

12. Simplify the expression.

$$(7a^3b^2 - 2a^2b + 5a^3b^2) - (3a^2b - 4a^3b^2)$$

A. $10a^8b^5 - 7a^5b^3$
B. $8a^3b^2 - a^2b$
C. $16a^3b^2 - 5a^2b$
D. $8a^3b^2 + 5a^2b$

13. Elsa deposited 20% of her babysitting earnings in the bank, which was $40. How much did she earn?

A. $60
B. $80
C. $100
D. $200

14. Which diagram best shows that angles x and y are supplementary?

A.

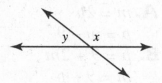

B.

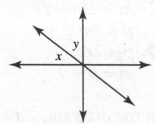

C.

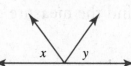

D.

15. Multiply the binomials below.
$(g - 6)(g + 3)$

A. $2g - 3$
B. $g^2 - 3g - 18$
C. $g^2 + 3g - 18$
D. $g^2 - 3$

16. Peter and Mike belong to the same scout troop. Peter has twice as many badges as Mike. Mike has 9 fewer than Peter. Which pair of equations can be used to determine the number of badges the boys have?

A. $m = 2p$
$p = m - 9$
B. $p = 9 + 2m$
$m = 9 - p$
C. $m = 2(p - 9)$
$p = 9m$
D. $p = 2m$
$m = p - 9$

17. In the diagram, $\angle a = 30°$. Find the measure of $\angle x$.

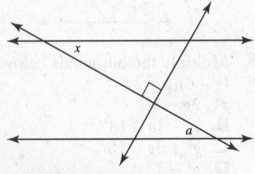

A. $30°$
B. $60°$
C. $100°$
D. $120°$

18. Simplify.
$(2x + 7x - 4) - (9x + 6 - 3x)$
A. $3x + 2$
B. $3x - 10$
C. $15x + 2$
D. $15x - 10$

19. The Smith family is adding a rectangular deck to their home as shown in the diagram. The carpenters make sure the deck has square corners by measuring the diagonals. If $AB = CD = 25$ feet and the length of the deck is 20 feet, how wide is the deck?

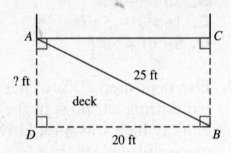

A. 30 ft
B. 20 ft
C. 15 ft
D. 12 ft

20. Simplify.
$2x(x + 3) - x(x - 2)$
A. $3x^2 - 6x$
B. $x^2 + 8x$
C. $2x^2 + 5x$
D. $x^2 + 4x$

21. A number, when added to 12, is equal to the number squared. Which equation represents this relationship?

A. $t^2 = t + 12$
B. $(t + 12)^2 = t - 12$
C. $t^2 + 12 = t$
D. $t = 12 - t^2$

22. After $2\frac{1}{2}$ years at an interest rate of 3%, your investment earned $15. What amount did you deposit?

A. $200
B. $450
C. $2,000
D. $4,500

23. $\angle A = 5x + 9$ and $\angle B = 3x + 11$. What is the measure of $\angle B$?

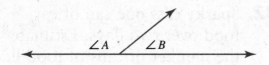

A. 20°
B. 25°
C. 65°
D. 71°

24. Jill made a 15% commission on every piano she sold. Last month her sales total was $30,000. What did she make in commissions?

A. $4,500
B. $450
C. $400
D. $45

25. Find the measure of $\angle x$.

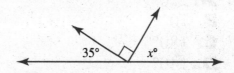

A. 35°
B. 55°
C. 20°
D. 145°

26. A bike, originally priced at $200, is now on sale for $160. What percent represents the savings?

A. 60%
B. 40%
C. 10%
D. 20%

27. Two parallel lines are cut by a transversal. If $\angle b = 110°$, what is the sum of the measures of $\angle c$ and $\angle e$?

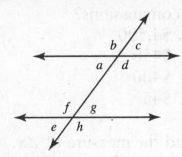

A. 220°
B. 70°
C. 180°
D. 140°

28. Factor this expression.

$$16x^6 - 8x^4 + 12x^3$$

A. $8(2x^6 - x^4 + 4x^3)$
B. $2x^3(8x^3 - 4x + 6)$
C. $x^3(16x^3 - 8x + 12)$
D. $4x^3(4x^3 - 2x + 3)$

29. Which is the best bargain for a bag of gourmet cat food?
A. 1 lb bag for $3.29
B. 2 lb bag for $6.29
C. 5 lb bag for $14.45
D. 8 lb bag for $25.52

30. Which choice best represents the simplified expression below?

$$(2x^3)^3 + (x^4)^2$$

A. $8x^9 + x^8$
B. $2x^{12}$
C. $2x^9 + x^8$
D. $7x^6$

31. Shari has saved $12 to buy a pair of $59 boots. While working as a tutor, she will earn $8 an hour. Which inequality represents the number of hours she must work to have enough money to buy the boots?
A. $12x + 8 \geq 59$
B. $8x - 59 \geq 12$
C. $8x + 12 \geq 59$
D. $12x - 8 \geq 59$

32. Sparky eats one can of cat food over two days. Estimate the number of cans of food he will eat in four months.
A. 15
B. 60
C. 120
D. 240

33. Simplify the expression below.

$$5x^2 + 4x + 2x^2 - 4x + 3$$

  A. $7x^2 + 4x + 3$
  B. $7x^2 + 3$
  C. $7x^2 + 8x + 3$
  D. $7x^2 + 3x$

34. The hypotenuse of a right triangle is 15 meters. Find the length of the leg if the base is 12 meters.

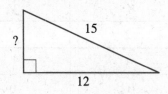

Not drawn to scale

  A. 3
  B. 9
  C. 12
  D. 30

35. If the cost of apples is $4.79 for a 3-pound bag, estimate the cost of 10 pounds of apples at the same rate.

  A. $12
  B. $14
  C. $16
  D. $48

36. Evaluate the expression $8x^3 - 2x^2 + 4x$ when $x = 2$.

  A. 20
  B. 36
  C. 48
  D. 64

37. The distance between Hilldale and Armorville measures $2\frac{3}{4}$ inches on a map where the scale is 1 inch = 50 miles. What is the distance between the 2 cities to the nearest mile?

  A. 125.75
  B. 135
  C. 137.5
  D. 138

38. A sweater that costs $28 is on sale at 20% off. Terri has a coupon for an additional 10% off. Which expression can be used to find the reduced price for the sweater?

  A. $(28)(0.8)(0.9)$
  B. $28 - 0.2 - 0.1$
  C. $(28)(0.2)(0.1)$
  D. $28 - 0.8 - 0.9$

39. If $A = 2x$ and $B = 3$, which expression represents $(A + B)^2$?

  A. $4x + 9$
  B. $2x^2 + 9$
  C. $4x^2 + 12x + 9$
  D. $2x^2 + 6x + 9$

40. Which transformation is shown on the coordinate plane below?

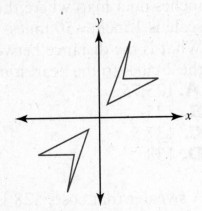

- **A.** translation
- **B.** dilation
- **C.** reflection
- **D.** rotation

41. Which is the coefficient of the term $5x^3$?
- **A.** 5
- **B.** x
- **C.** 3
- **D.** x^3

42. Choose the equation of a line that passes through the points (0, 6) and (8, 10).
- **A.** $y = 2x + 6$
- **B.** $y = \frac{1}{2}x + 6$
- **C.** $y = 2x + 10$
- **D.** $y = \frac{1}{2}x + 8$

PART 2

Directions

You have 70 minutes to answer the following extended-response questions. Be sure to explain your solutions clearly. Write legibly and neatly. You may use your calculator to solve these questions.

1. The table below shows a function rule.

x	y
–2	1
–1	
0	0
1	–0.5
2	
3	–1.5
4	

- **A.** Fill in the missing numbers.
- **B.** Write the function rule based on the table.

2. Figure *PQRS* is a rhombus created from two sets of parallel lines. The diagonal *QS* divides the rhombus into two congruent triangles. If ∠*PQS* = 65°, find the measure of ∠*QPS*. Show your work.

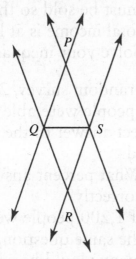

3. If $y = \dfrac{x-7}{4}$ and $y = 9$, find the value of *x*. Show your work.

4. A vegetable garden is fertilized once a week. Eight ounces of a granular fertilizer are mixed with 5 gallons of water and then sprayed on the plants. Approximately how many months will a 20-pound bag of the granular fertilizer last? Explain how you arrived at your answer.

5. Rotate *ABCD* counterclockwise 90° from the center of rotation at (0, 0). Label the image *A'B'C'D'*.

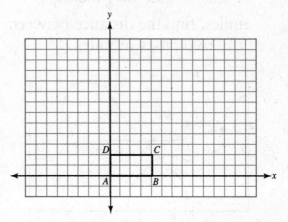

6. In the diagram below, lines *c* and *d* are parallel. Triangle *JKL* is a right triangle. Find the measure of ∠*t*.

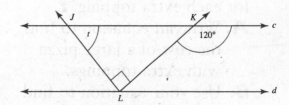

7. Mr. Townsend is the director of the afterschool program. He is planning a bus trip to the museum. Sixty-five percent of the students in the program filled every seat on 3 buses. Each bus has 52 seats. How many students belong to the afterschool program?

8. The distance between Harmony and Greenville is 3.5 inches on the map below. Using a scale of $\frac{1}{4}$ inch = 15 miles, find the distance between the two cities in miles.

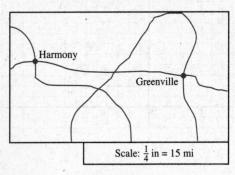

Scale: $\frac{1}{4}$ in = 15 mi

Not drawn to scale

9. The cost of a large pizza with cheese, p, is $7.50 plus $1.50 for each extra topping, t.
   A. Write an equation to find the cost of a large pizza with extra toppings.
   B. Use your equation to find the cost of a pizza with three extra toppings.

10. The bookstore sells pencils at $0.25 and markers at $0.70 each. So far, all the pencils have been sold, and the income from their sale is $28.
   A. Write an inequality to find the number of markers that must be sold so that the total income is at least $80.
   B. Solve your inequality.

11. In a random survey, 24 out of 150 people were able to give a correct answer to the question asked.
   A. What percent answered correctly?
   B. If 1,200 people were asked the same question, how many would be expected to answer correctly?

12. Two parallel lines are cut by a transversal as shown below. $\angle a = 2(6x - 1)$ and $\angle e = 3x + 2$.

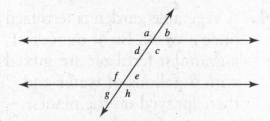

   A. Find the measures of the following angles.
   $\angle d = $ _____
   $\angle f = $ _____
   $\angle g = $ _____
   B. Name one pair of vertical angles.

Answers to Sample Test 2

The solutions presented here for this sample test are not the only possible solutions. You may know another way to solve the problem and arrive at the correct answer. Looking for different ways to solve the same problem helps improve your problem-solving skills.

PART 1

1. D	10. B	19. C	28. D	37. D
2. B	11. A	20. B	29. C	38. A
3. A	12. C	21. A	30. A	39. C
4. A	13. D	22. A	31. C	40. D
5. D	14. B	23. D	32. B	41. A
6. C	15. B	24. A	33. B	42. B
7. D	16. D	25. B	34. B	
8. A	17. A	26. D	35. C	
9. C	18. B	27. D	36. D	

1. (D) $x - 3$

$$\frac{3x^3 - 9x^2}{3x^2} =$$

$$\frac{3x^3}{3x^2} - \frac{9x^2}{3x^2} =$$

$$x - 3$$

2. (B) $\angle 1$ and $\angle 4$

Angles 3 and 4 are vertical angles and are congruent. Angles 1 and 3 are corresponding angles. All three angles are equal in measure.

3. (A) 54

$2x^3 = 2(3^3) = 2(27) = 54$

Simplify exponents before multiplying, following the order of operations.

4. (A) $11k^2m^3$

These are like terms. Just add the coefficients.

5. (D) $2x \geq 75$

The students make $8 – $6 or $2 profit on each plant. Their profit must be equal to or more than $75.

6. (C) $(y + 2)$ and $(y - 6)$

Two factors of –12 that add up to –4 are +2 and –6.

7. (D) $180 - 3x$

The sum of supplementary angles is 180°. Subtract $3x$ from 180 to find the measure of $\angle YXZ$.

8. (A) The number of guppies is 4 more than 3 times the number of mollies. When written as equations, Choice B is $g = m - 8$, Choice C is $g = 4m + 3$, and Choice D is $g = 3m - 4$.

9. (C) $6x^2 + 7x - 20$

Using FOIL, we have $2x(3x) + 2x(-4) + 5(3x) + 5(-4)$

$6x^2 - 8x + 15x - 20$

$6x^2 + 7x - 20$

10. (B) $\angle 1$ and $\angle 4$ are congruent.

For Choice A, $\angle 2$ and $\angle 3$ are supplementary. For Choice C, $\angle 2$ and $\angle 4$ are congruent. For Choice D, $\angle 1$ and $\angle 3$ are supplementary.

11. (A) $200 = 1.15s$

If 15% of the seeds never grow, then 15% more have to be planted.

So s + 15% more seeds should be planted $\Rightarrow s + 0.15s \Rightarrow 1.15s$

12. (C) $16a^3b^2 - 5a^2b$

$(7a^3b^2 - 2a^2b + 5a^3b^2) - (3a^2b - 4a^3b^2) =$
$7a^3b^2 - 2a^2b + 5a^3b^2 - 3a^2b + 4a^3b^2 =$
$7a^3b^2 + 5a^3b^2 + 4a^3b^2 - 2a^2b - 3a^2b =$
$16a^3b^2 - 5a^2b$

13. (D) $200

One way to solve this problem is with an equation.

20% of $x = 40$
$0.2x = 40$
$0.2x(10) = 40(10)$
$2x = 400$
$x = 200$

Calculators are not allowed on this portion of the test. Multiplying both sides of the equation by 10 moves the decimal point.

You can also solve this problem using a proportion.

$$\frac{20}{100} = \frac{40}{x}$$

$20x = 4,000$
$x = 200$

14. (B)

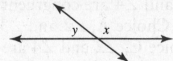

Choice A shows congruent angles. Choice C shows complementary angles. Choice D shows nothing specific.

15. (B) $g^2 - 3g - 18$
With FOIL, $(g - 6)(g + 3) =$
$g^2 + 3g - 6g - 18 =$
$g^2 - 3g - 18$

16. (D) $p = 2m$ and $m = p - 9$
Let p represent Peter's badges, and let m represent Mike's badges.

17. (A) 30°
Angles a and x are alternate interior angles and are congruent.

18. (B) $3x - 10$
$(2x + 7x - 4) - (9x + 6 - 3x) =$
$2x + 7x - 4 - 9x - 6 + 3x =$
$2x + 7x - 9x + 3x - 4 - 6 =$
$3x - 10$

19. (C) 15 ft
Use the Pythagorean theorem to solve for the side of the triangle.
$a^2 + b^2 = c^2$
$a^2 + 20^2 = 25^2$
$a^2 + 400 = 625$
$a^2 = 625 - 400$
$a^2 = 225$
$a = 15$
Notice that 20 and 25 are multiples of 5. This is a 3-4-5 triangle whose sides are multiples of 4, or 15-20-25.

20. (B) $x^2 + 8x$
$2x(x + 3) - x(x - 2) =$
$2x^2 + 6x - x^2 + 2x =$
$2x^2 - x^2 + 6x + 2x =$
$x^2 + 8x$

21. (A) $t^2 = t + 12$

22. (A) $200
Substitute the known values into the formula.
$$i = prt$$
$$15 = p(0.03)(2.5)$$
$$15 = 0.075p$$
$$15000 = 75p$$
$$200 = p$$

23. (D) 71°
$$m\angle A + m\angle B = 180°$$
$$5x + 9 + 3x + 11 = 180$$
$$8x + 20 = 180$$
$$8x = 160$$
$$x = 20$$
Substitute 20 in $3x + 11$ to find the measure
of $\angle B$, which is $60 + 11 = 71°$.

24. (A) $4,500
10% of $30,000 = $3,000
Since Jill made more than 10%, the only possible
answer is $4,500.

25. (B) 55°
The three angles compose a straight angle measuring 180°. You know that the right angle is 90° and
that the sum of the other two angles is 90°.
$$90° - 35° = 55°$$

26. (D) 20%
The savings is $40. One way to solve this problem
is with a proportion.
$$\frac{x}{100} = \frac{40}{200}$$
$$200x = 4,000$$
$$x = 20$$

27. (D) 140°

Angles b and c are supplementary, which makes $\angle c = 70°$. Angles c and e are congruent, so their sum is 2(70) or 140°.

28. (D) $4x^3(4x^3 - 2x + 3)$

The GCF is $4x^3$. Divide each term by the GCF.

$$\frac{16x^6 - 8x^4 + 12x^3}{4x^3} = 4x^3 - 2x + 3$$

29. (C) 5 lb bag for $14.45

If a 1 lb bag costs $3.30, then a 2 lb bag should cost $6.60, a 5 lb bag should cost $16.50, and an 8 lb bag should cost $26.40. The largest difference between the costs listed above and the choices is (C).

30. (A) $8x^9 + x^8$

Use the rules of exponents.
$(2x^3)^3 + (x^4)^2 =$
$(2^3)(x^3)^3 + (x^4)^2 =$
$8x^9 + x^8$

31. (C) $8x + 12 \geq 59$

Choose the equation that shows $8 times the number of hours worked (x) plus Shari's savings ($12) should be greater than or equal to the cost of the boots ($59).

32. (B) 60

Estimate the number of days in a month at 30. This means Sparky eats 15 cans per month. In four months he will eat (15)(4) or 60 cans.

33. (B) $7x^2 + 3$

Combine the like terms.
$5x^2 + 4x + 2x^2 - 4x + 3 =$
$(5x^2 + 2x^2) + (4x - 4x) + 3 =$
$7x^2 + 3$

34. (B) 9

The lengths of the sides of this triangle are multiples of the 3-4-5 Pythagorean triple. Multiply each number by 3 to get a 9-12-15 triangle. The missing side length is 9.

35. (C) $16

Let's say the 3-pound bag costs $4.80. That means a 1-pound bag costs $4.80 ÷ 3 or $1.60. A 10-pound bag will cost ($1.60)(10) or $16.

36. (D) 64

Substitute 2 for x.
$$8x^3 - 2x^2 + 4x =$$
$$8(2^3) - 2(2^2) + 4(2) =$$
$$8(8) - 2(4) + 4(2) =$$
$$64 - 8 + 8 = 64$$

37. (D) 138

$(2.75)(50) = 137.5 \approx 138$

Notice that the question asked for the distance to the nearest mile. So round up the decimal answer.

38. (A) (28)(0.8)(0.9)

Remember that multiplication is associative, so the order in which you multiply does not matter. Also remember that when an item is on sale for 20% off, you pay 80%. Instead of doing more steps than needed, use the shortcut by multiplying the cost of the sweater by 0.8 and again by 0.9.

39. (C) $4x^2 + 12x + 9$

Substitute and then multiply.
$$(A + B)^2 =$$
$$(2x + 3)^2 =$$
$$(2x + 3)(2x + 3) =$$
$$4x^2 + 6x + 6x + 9 =$$
$$4x^2 + 12x + 9$$

40. (D) rotation

Notice that the back points of the arrow have different thicknesses. The shape was rotated.

41. (A) 5

The coefficient is the number multiplied by the term. In this term, 5 is the coefficient. The exponent is 3.

42. (B) $y = \dfrac{1}{2}x + 6$

From looking at the coordinates of the points, you can see that the y-intercept is 6. Choices (C) and (D) are eliminated. Find the slope using the formula.

$$m = \frac{y_2 - y_1}{x_2 - x_1}$$

$$\frac{10 - 6}{8 - 0} = \frac{4}{8} = \frac{1}{2}$$

Substitute the slope and y-intercept into the slope-intercept form of a linear equation.

$$y = \frac{1}{2}x + 6$$

PART 2

1. A. The missing values are 0.5, –1, and –2.

   B. The value of y is found by dividing the value of x by –2.

$$y = \frac{-x}{2} \text{ or } x = -2y$$

2. Since the two triangles are congruent, $\angle PQR$ is $2(65°)$ or 130°. Angles RQP and QPS are supplementary. So $\angle QPS$ is $180° - 130°$ or 50°.

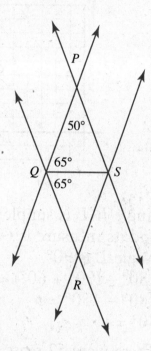

3. Substitute 9 for y in the equation and then solve for x.

$$\frac{9}{1} = \frac{x-7}{4}$$

$$36 = x - 7$$

$$36 + 7 = x - 7 + 7$$

$$43 = x$$

4. Eight ounces is $\frac{1}{2}$ lb. One pound is used every 2 weeks. Then 20 lb will last 40 weeks. Since there are about 4 weeks in a month, the bag will last approximately 10 months.

5. A 90° rotation turns the rectangle a quarter-turn to the left.

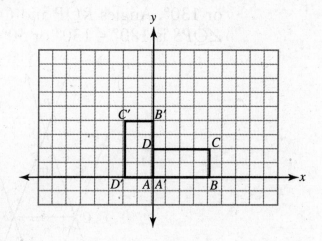

6. Angle *JKL* is supplementary to the given 120° angle, and its measure is 60°. There are 180° in a triangle. Angle *L* is 90°.

$$180° - (90° + 60°) = t$$
$$180° - 150° = t$$
$$30° = t$$

7. There were 52 seats on each of 3 buses, or 156 students who went on the trip.

$$\frac{65}{100} = \frac{156 \text{ students on trip}}{x \text{ members}}$$
$$65x = 15{,}600$$
$$x = 240$$

There are 240 students in the afterschool program.

8. $$\frac{0.25 \text{ in.}}{15 \text{ mi}} = \frac{3.5 \text{ in.}}{x \text{ mi}}$$

$$0.25x = 52.5$$

$$x = 210$$

The distance between the cities is 210 miles.

9. A. $p = 7.5 + 1.5t$

B. Substitute 3 for t in your equation.

$p = 7.5 + 1.5t$

$p = 7.5 + 1.5(3)$

$p = 7.5 + 4.5$

$p = 12$

The pizza will cost $12.

10. A. Amount from pencils plus amount from markers $\geq \$80$.

$28 + 0.7m \geq 80$

B. $28 + 0.7m \geq 80$

$28 - 28 + 0.7m \geq 80 - 28$

$0.7m \geq 52$

$m \geq 74.285$

75 markers must be sold.

11. A. $$\frac{24}{150} = \frac{x}{100}$$

$$2,400 = 150x$$

$$16 = x$$

16% of the people answered correctly.

B. 16% of 1,200 =

$0.16(1,200) = 192$

If asked, 192 people would be expected to answer correctly.

12. Angles a and e are supplementary.

$2(6x - 1) + 3x + 2 = 180$

$12x - 2 + 3x + 2 = 180$

$12x + 3x - 2 + 2 = 180$

$15x = 180$

$x = 12$

Substitute 12 for x in the expressions to find the measure of one of the angles.

$\angle a = 2(6x - 1) = 12x - 2 =$

$12(12) - 2 =$

$144 - 2 =$

$142°$

$\angle e = 3x + 2 = 3(12) + 2 = 36 + 2 = 38°$

A. Angles b, d, e, and g are 38°. Angles a, c, f, and h are 142°.

B. There are four pairs of vertical angles: a and c, b and d, e and g, and f and h.

Index

A

Absolute value, 10
Acute angles, 82
Addition
  description of, 9
  of fractions, 14–16
  of integers, 11
  order of operation for, 13
  of polynomials, 65
Additive inverses, 7, 9, 31
Algebraic expressions. *See*
    Expressions
Alternate exterior angles, 85
Alternate interior angles, 84
Angles, 82–85, 95
Associative property, 9, 31
$ax^2 + bx + c$, 78–79

B

Base, 20
Binomials
  definition of, 65
  multiplication of
    by a binomial, 72–74
    by a monomial, 69–70
  squaring, 74–75

C

Celsius, 60–61
Circle, 143
Coefficient, 30
Commutative property, 9, 31
Complementary angles, 83
Composite numbers, 20
Congruence, 95–97
Congruent angles, 84, 92, 95
Coordinate system, 103
Corresponding angles, 84
Counting numbers, 7
Cross products, 42
Customary system of measurement,
    57–58, 60

D

Decimals
  changing of, to percents, 46
  fractions changed to, 45–46
Denominator, 45
Dependent variables, 125
Descartes, René, 103

Descriptive graphs, 129–131
Difference of two perfect squares, 74
Dilation, 150–152
Distributive over addition property,
    9, 71, 76
Divisibility rules, 18
Division
  of exponents, 67–68
  of integers, 12–13
  order of operation for, 13
Domain, 125

E

Equations
  definition of, 30
  of horizontal lines, 111–112
  of lines, 115–117
  quadratic, 122–124
  solving of, 30–32, 117–119
  of vertical lines, 111–112
  writing, 34–37
Equilateral triangle, 85, 91
Exponents
  definition of, 13, 20
  division of, 67–68
  law of, 66–68
  multiplication of, 67
Expressions
  evaluating, 25–26
  factoring, 76–77
  writing, 23–29

F

Fahrenheit, 60–61
FOIL method, 72, 78
Fractions
  addition of, 14–16
  changing of
    to decimals, 45–46
    to percents, 47
  improper, 16–17, 45
  multiplication of, 16–18
  subtraction, 14–18

G

Geometry
  angles, 82–85
  lines, 81–82
  polygons, 91–95
  quadrilaterals, 90, 92

  triangles. *See* Triangles
Graphing
  coordinate system, 103
  inequalities, 32–33, 120–122
  lines, 107–117
  plotting points, 104
  slope, 105–107
Greatest common factor (GCF), 19,
    76–77

H

Hexagon, 93
Horizontal lines, 111–112
Hypotenuse, 87–89

I

Identity element for multiplication,
    10
Improper fractions, 16–17, 45
Independent variables, 125
Inequalities
  definition of, 30
  graphing, 32–33, 120–122
  linear, 120–122
  solving, 32–33
  writing, 34–37
Integers
  adding of, 11
  definition of, 7
  dividing of, 12–13
  multiplying of, 12
  subtracting of, 11
Interest, 52–53
Interior angles, 92–95
Irrational numbers, 8
Isosceles trapezoid, 90
Isosceles triangle, 85–86

L

Law of exponents, 66–68
Least common denominator (LCD),
    15
Least common multiple (LCM), 19
Like terms, 30
Line(s)
  graphing, 107–117
  slope of, 105–107
  types of, 81–82
  writing equations of, 115–117
Linear equation, 107–108